Impossible Visits

Also by the Author

Hazard and the Five Delights (novel)
In the Unlikely Event of a Water Landing: A Geography of Grief (memoir)
The Sea Monkey Tombs (novel)
A Frail House (short stories)

Impossible Visits

The Inside Story of Interactions with Sasquatch at Habituation Sites

Christopher Noël

ISBN: Softcover 978-1-4363-9851-0

This book was printed in the United States of America.

To order additional copies of this book, contact:
Xlibris Corporation
1-888-795-4274
www.Xlibris.com
Orders@Xlibris.com
57814

CONTENTS

STATEMENT OF PURPOSE AND CALL FOR FURTHER STORIES

Time is running out. Before long, a member of the Sasquatch race will be shot and killed, or struck down by a late-night vehicle, or a corpse will be stumbled upon in the wilderness. I know of dozens of men currently on the hunt in North America, aggressively, single-mindedly, and with high-tech weaponry; large bounties are waiting to be claimed.

When the day arrives that a body is hauled before the bright lights of the media and a flabbergasted public, a great need will suddenly arise for reliable facts, and therefore we must, at that moment, already have in place a solid infrastructure of knowledge. Otherwise, the vacuum will be filled with misinformation and ill intent.

This is where the habituators come in, women and men who have been interacting with these fascinating beings, in some cases for decades, and carefully documenting their work, in the tradition of Jane Goodall and Dian Fossey. I myself am a newcomer to the habituation process, but have met with preliminary success, and have been privileged to connect with and learn from those who have gone before me. All the knowledge and understanding that they have gathered, patiently and quietly over the years, will likely be the only effective barricade, post-"discovery," standing between Sasquatch and the worst-case scenario: an all-out open season, leading toward genocide. The more generous the habituators throughout the continent are able to be in sharing what they have found out, the more informed the populace at large will become, and the more trouble the killers will have in forcing their agenda.

The public revelation will be, of course, a pivotal moment in history, and the fate of this harmless species will hang in the balance. I and the contributors to the book and DVD seek to place the concrete and subtle reality of this advanced race into a respectful, loving context. Our intent is to frame the matter clearly enough, and in enough convincing detail, to counteract the false and sensationalistic interpretations that will spread over the land like wildfire. Otherwise, by default, the media will seek ratings and profit by scaring people.

It is certain that enough Sasquatch/human interaction has taken place, and continues to take place, to fill many books and DVDs. I would be honored to gather as much of this material together as I am able, within subsequent editions of *Impossible Visits*. If anyone out there has had ongoing experiences with these complex, subtle and magnificent creatures, or knows of such cases, I invite you to contact me directly at *impossiblevisits@aol.com*.

And please know that I will treat each habituator, each testimonial, and all information shared, with the same strict respect for anonymity, and care for the concealment of geographical specifics, that I have employed in the present volume.

Thank you sincerely.

IMPOSSIBLE VISITS: THE DVD

This 110-minute documentary opens windows into the work of the habituators who have contributed to the book, chronicles my "Vermont Ravine Project: 2007 & 2008," and includes scenes from BFRO expeditions. Samples from the DVD may be viewed at www.impossiblevisits.com. In addition to their own voices recounting their experiences, and footage of habituators' research sites, this documentary includes never-before-released photographs and video of their Sasquatch visitors.

For instance, "A Kansas Gallery" features the long-distance photography of a researcher who prefers, at this point, not to speak publicly about his work. Quietly and patiently, he has become perhaps the finest Sasquatch photographer in the world today; one of his images graces the back cover of this book.

None of the visual material will rise to the level of "proof" in the eyes of holdout skeptics. Neither the book nor the DVD sets out to explore *whether* this species exists (a stale, bygone debate), but, rather, *how*.

PREFACE

> His power, they know, resides in his absence, a continuous absence that exaggerates the imagination of these persons who for centuries have been calling themselves, simply, his children.
>
> —novelist Kate Walbert

The first time I saw a Sasquatch was in Texas, on the evening of November 8, 2008, and I didn't even know it. One of the researchers contributing to this book had made an impressive bonfire in her backyard. She and I, her twenty-year-old daughter, and her daughter's friend, were roasting hot dogs and marshmallows, and just goofing around.

From time to time, I'd leave the fire area and scan with the thermal imager (a machine that reads "heat signatures") along the nearby tree line. Yes, I did notice the vague bar of light (heat) at the periphery, low and parallel to the ground, at the edge of the firelight, but for some reason I didn't think it worthy of a closer approach. In retrospect, of course, I wish I had, but after all, I'd geared myself to pay attention to sudden upright giants, not to anything so small and horizontal.

It's not until nine days later, back home in Vermont, that I review the footage and realize that the bar of light is actually *moving*, and in a peculiar way, tough to interpret. The front end keeps thrusting forward and down, to the right. I assume this object on my computer monitor must be a hog or a sheep, or *something* ordinary in the woods, and that this front end is its head, perhaps choking or making a call that we couldn't hear, thanks to the loud bonfire.

I enhance the footage, upping the contrast, and send the file around to the habituators' group, people who have had repeat visitations to their properties. The Oklahoma woman says the magic word; she's noticed that as the "head" moves, it *splays,* like the fingers of a hand, flexing. I look again, and again. In an abrupt gestalt shift, the true nature of the image jumps out at me plainly. This long, bright object is, in fact, a left arm and a hand, slung over a dark, pointed board, the hand actually cupping the tip of the board at times, the rest of the figure crouched and hidden behind a heap of debris.

Three times, the Sasquatch stretches its hand like this, out and down, like someone with a cramp. If it hadn't moved, no identification would have been possible. Perhaps it's signaling to someone on the ground behind it: "Stay down! Stay down!"

A little later, after ignoring the marvel that peers at us just fifty feet from our chairs, after eating hot dogs and bantering with the others, I aimed back at the bar of light, as at some bland consolation prize, and this time (on the computer) I find that it remains still, the hand folded now into a fist. Knowing now what it is I'm looking at, I can clearly see that the fist shines brighter (giving off more heat) than the forearm, because of the difference in hair-cover.

I grew up in a village of 450, in Vermont. A mile away from our house, there is a ravine that I liked to explore. It's a steep ravine, more than a thousand feet deep, maybe a quarter mile wide and a mile long, undeveloped back then, undeveloped still today. When I was a teenager, I felt drawn to it, as to some otherworldly, ancient spot on earth. The air smelled different inside there, as though it was space without time.

And now, thirty-three years later, I've spent hundreds of hours in this place, and many overnights, and never encountered a fellow

Homo sapien. I have, however, encountered Sasquatch, in various manifestations, both subtle and otherwise—see Chapters Three and Four. If you're down inside the ravine, in your sleeping bag at three o'clock in the morning, human civilization seems a pale fiction.

When I'd explore the woods behind our house, as a kid, I'd peek in holes and hollow trees, scan carefully along ridge lines, positively aching to catch sight of a Hobbit, a gnome, a hulking ogre, or some such utterly *other* kind of creature than us, non-human yet also *like* us, frisky and sympathetic. The forest as a whole resembled a house much larger on the inside than on the outside, rewarding belief, unfolding room by room, pocket by secret pocket.

The idea of Sasquatch first reared up before me, as for most of us, thanks to the 1967 Patterson/Gimlin Film. Of course, I wanted to travel there—to the Pacific Northwest—immediately, to where this creature actually *lived,* and over the years, this vision persisted, imprinted and always astir, serving me in the way of an elusive deity, like the beckoning horizon, lush because unattainable.

One snowy day in early 1992, my fiancée was killed in a car accident. Seeking comfort and transcendence, I decided it was high time to treat myself to a solo expedition to British Columbia, to the little town of Bella Coola, which had logged more than its share of Sasquatch sightings and track finds over the years. I flew to Vancouver, rented a mini-van, and drove the hundred and forty-five miles due north. Unfortunately, I was so afraid of grizzly bears that I didn't dare venture more than fifty yards from my vehicle. Instead, I put in hundreds of miles driving along narrow forest service roads and interviewed locals who claimed to have seen the creature.

They were credible and understated, these townsfolk. Actually, one young man didn't even claim this much but said he'd seen a leg, just a leg—white, furry, thick—of something walking upright in his backyard, the rest of the body obscured by darkness and fog. He asserted several times, "I'm not saying I know what it was. I saw *something.*"

Ten years later, my father died suddenly of a heart attack. A scholar and writer, he'd been a life-long champion of the imagination, of *my* imagination, in the most robust sense of the word, not reduced to the merely fanciful but endowed with the potency of true insight: a window showing vistas beyond the easy consensus reality we've all been taught.

This death, too, sent me back to the long dream. On-line, I found the Bigfoot Field Researchers Organization, a group that had been actively studying the Sasquatch phenomenon for years. I joined up, and attended ten expeditions in far-flung regions of North America, traveling more than nineteen thousand miles. I met like-minded men and women, also years along in this seeking.

My childhood prayer of a forest housing upright *others* finally stopped feeling so childish. The very ground itself now seemed renewed, re-enchanted, thanks to nothing more than five-toed footsteps.

Still, the desire: If only I could actually *live* in one of these Sasquatch "hot spots."

Fourteen years after losing my fiancée, I fell in love again, and together, she and I brought a girl child into the world. At forty-five, I became a first-time father.

The break-up felt like another death.

Twenty-four days later, a report came into the BFRO Web site suggesting Sasquatch activity just eighteen miles north of me. To say the least, pursuing this case provided a welcome distraction. This man, Ojibwa by descent, turned out to be highly credible, his evidence compelling, and learning from him opened my eyes to this heretofore-inconceivable possibility of *local* Sasquatch. I came to understand that the primate species inhabited—at least occasionally—the very ravine I'd visited as a boy, and that is not far from my front door even today.

I'm not suggesting any mystical, causal relationship between these three deaths and my approach to this creature; but I do think that they have thrown me open, at some level, to what lies beyond the human, yet offers kinship.

Yet, there's more to it, as well. Indeed, sometimes people—dead *or* alive—simply are not enough. Nor is the known world as a whole. We seem to retain some level of grief even in the best of times, some hollow in the heart, some longing to pass through to another world than this one. After all, what's available for us to *be with*, here? Three categories: the human, the animal, and the divine. Sasquatch fits into none of these, exhibiting qualities of all three, yet opening up a fourth type of place.

If the Judeo-Christian divine is too abstract, as it is for me, then the mind may turn to notions from mythology. This is what Sasquatch reminds me of: the gods of Mount Olympus. They are part of nature yet stand beyond nature, too: they cannot be circumscribed and contained. They seem to travel wherever they will, like quicksilver, otherwise known as Mercury, otherwise known as Hermes, fleet of foot, messenger between realms, shape-shifter, trickster, dweller on the threshold. They occupy their own dimension, yet occasionally visit our own, when it suits their purpose, to give us clues, glimpses, gifts, to play with us their oblique and capricious games. They are profoundly and irreducibly Other, yet many (as featured in this book) find that in the grace of their presence, they feel more themselves, in an expanded sense of Self. Why else would researchers spend decades stomping through wilderness, straining after even a glimpse? Many compromise their home lives in hopes of contact, of a moment of ecstasy—"ek-stasis": to stand outside oneself.

One of the principal figures in this book describes those moments when you cross an invisible border in the woods, when you suddenly get that *sizzle*: "You know," he says, "the sense that I've been here before . . . but I haven't?" It's the *déjà vu* that lets you into a new, old territory, resting on home ground for the first time.

It's as though you're coming face to face with what you once knew, perhaps eons ago, a recollection of the whole story.

Having finally, now, seen a Sasquatch—well, its wan, breathtaking appendage—I can report that the overriding feeling-tone is *relief.* A sudden saturation.

What has happened for me during these past three years, in researching people's accounts—more amazing, I think, more *quickening,* than most stories from literature—and in experiencing my own small fortune of direct contact, has been a grand odyssey.

A good friend once warned me, seeking to curb this folly of mine, "Don't forget, this is the only life you have."

I remember.

INTRODUCTION

The Role of "Habituation" Within the History of Primate Research

Jane Goodall's life was changed forever in 1960 when, in Nairobi, she experienced her first close encounter.

> For over half a year I had been trying to overcome the chimpanzees' inherent fear of me, the fear that made them vanish into the undergrowth whenever I approached. At first they had fled even when I was as far away as five hundred yards and on the other side of the ravine . . . [But now,] less than twenty yards away from me two male chimpanzees were sitting on the ground staring at me intently. Scarcely breathing, I waited for the sudden panic-stricken flight that normally followed a surprise encounter between myself and the chimpanzees at close quarters. But nothing of the sort happened. The two large chimps simply continued to gaze at me. Very slowly I sat down, and after a few more moments, the two calmly began to groom one another. (*In the Shadow of Man*)

Three years later, having recently set up a gorilla research station in the mountains of Rwanda, Dian Fossey experienced an initiation, as well.

> I shall never forget my first encounter with the gorillas. Sound preceded sight. Odor preceded sound in the form of an overwhelming musky-barnyard, humanlike scent. The air was suddenly rent by a high-pitched series of screams followed by the rhythmic rondo of sharp *pok-pok* chestbeats from a great silverback male obscured behind what seemed an impenetrable wall of vegetation . . . . Most of the females had fled with their infants to the rear of the group, leaving the silverback leader and some younger males in the foreground, standing with compressed lips. (*Gorillas in the Mist*)

"During the early days of the study at Kabara," writes Fossey,

> it was difficult to establish contacts because the gorillas were not habituated or accustomed to my presence and usually fled on seeing me. I could often choose between two different kinds of contacts: obscured, when the gorillas didn't know I was watching them, or open contacts, when they were aware of my presence.
>
> Fossey learned that the best way to habituate these creatures was simply to act like them.
>
> Open contacts . . . slowly helped me win the animals' acceptance. This was especially true when I learned that imitation of some of their ordinary activities such as scratching and feeding or copying their contentment vocalizations tended to put the animals at ease more rapidly than if I simply looked at them through binoculars while taking notes. I always wrapped vines around the

> binoculars in an attempt to disguise the potentially threatening glass eyes from the shy animals.

Jane Goodall, too, took a humble and consistent approach in order to familiarize her subject with her daily presence.

> Because I always looked the same, wearing similar dull-colored clothes, and never tried to follow them or harass them in any way, the shy chimpanzees began to realize at long last, that after all I was not so horrific and terrifying.

Famously, of course, through years of devoted patience, both Goodall and Fossey gained acceptance by their respective primate study groups, and it was this high threshold of immediacy and intimacy that allowed their research projects to break new ground.

Since the mid-1980s, researchers have found increasing success in making contact with Sasquatch, too, progressively refining their methods. For the most part, these methods have featured *acoustical* overtures—"wood knocks," "whoop calls," etc.—designed to elicit responses; when a colloquy does take place, it usually occurs late at night, quite often well past midnight. Like Goodall and Fossey, Sasquatch researchers, principally those of the Bigfoot Field Researchers Organization, have come to rely on such imitative behavior in order to play upon the animals' natural curiosity, and in the darkness, imitation means *sound*. Species interaction in this mode has now reached the point at which it can be achieved on a regular basis in many locations throughout North America. Regular, yes, and even predictable, yet by no means frequent. Perhaps five percent of the time, an overture will receive a recognizable response. But even this seemingly low figure far outweighs what was possible before.

Rarer still, during expeditions, sightings have taken place, an individual Sasquatch lured into view—again, mostly at night, and mostly very fleetingly, such as a head peeking from behind a tree—by these humans' oddly non-human behavior. Expedition-goers have caught glimpses through night vision infrared binoculars or scopes, but only quite recently has the gold standard of night-vision technology—hand-held thermal imagers able to record video—come within financial reach of some researchers or, more accurately, of generous sponsors of researchers; and still, a good one costs nine thousand dollars. Thermal imagers are such a gold standard, in this specialized area of primate research, because 1) they are an entirely *passive* instrument, emitting nothing, able to read the "heat signatures" given off by objects, especially living organic objects, whereas most other night-vision devices are *active,* giving off an infrared beam that can, it seems, be seen and avoided by this animal, and 2) thermal imagers can "see" even in the thickest dark, when the Sasquatch feels most comfortable and will often draw nearer.

As "shy" as Goodall and Fossey found their chimpanzees and mountain gorillas to be, Sasquatch is a hundred-fold more evasive, and smarter; otherwise, such a creature never would have been able to survive alongside human beings for several hundred millennia, and to avoid our increasingly weaponized domination.

Jane Goodall tells, for instance, of the chimp she called David Greybeard becoming brazen enough in time to simply stride into her camp and take bananas from her hand, his example soon persuading other members of his troupe to follow suit.

Dian Fossey, also, was eventually able to touch, even to groom and be groomed by, her subjects.

In the case of Sasquatch, however, this level of physical contact and interaction does not fall within the researchers' range of rewards.

At the same time, the Holy Grail of Sasquatch research does remain consistent habituation, this concept taken, though, in a more attenuated sense than in the case of other primates. The situations described in detail in this book represent the furthest advances toward genuine contact and interaction yet achieved. Or, to be more precise, these are our furthest *documented* advances, for it must be assumed that many times, over the vast stretch of human/Sasquatch co-evolution, interaction has occurred, some of it no doubt rising to the level of familiar regularity. But this must always have represented the rare exception, when, under the right circumstances, they will indulge in a highly textured—albeit still exceedingly cautious—degree of contact. While we may place the following examples in the category of "play," the implied intellect behind these gambits is much more nuanced than even the clever chimp or wise gorilla will manifest.

One of my neighbors in Vermont is a woodsman. Sylvester has been interacting with Sasquatch for years, though he has never laid eyes on one. That may sound like a naively credulous statement, until you realize that he has heard wood knocks and distinct—yet unintelligible—voices in the middle of nowhere, and found many of the simple stick and tree structures that have been reported throughout North America in close proximity to well-documented Sasquatch sightings. These pieces of evidence by themselves would not, I think, be compelling enough in Sylvester's case, were it not for the added layer of game-playing.

> I put two quarters on the ground and put them about a foot apart.
>
> Right out in plain sight. My idea is that no one could resist a couple quarters. If they disappeared then I know I have people in my woods and should keep an eye and ear open for them. (I fear people in the woods.) I also had them facing up and pointed them to face north. Nothing happened the first few years . . . till last year. Then a quarter was gone. I examined the area and found in the spot of the quarter, a cute little arrangement of feathers. Each a different color. And spread out facing east. If it were a person they would have taken both quarters and would not have left such a beautiful little arrangement.

> That is just like the games I played with the thing in the [Town Name] Woods. I would put sticks or stones next to the trail in various positions and they would be rearranged when I returned a few hours later. I am a very observant person, especially out in the woods. At one time it put a broken arrow into the ground so I took the two pieces and placed them on a stump next to where they had been. I faced them north. When I returned only one piece was there and the other piece was stuck in the ground again.

Hundreds of reports, continent-wide, of such quiet, sly pranksterism, share an undercurrent of strangeness, humor, and a kind of tacit empathy.

A fellow researcher in Upstate New York wrote me of an ongoing situation he experienced at his remote trailer, which culminated in the following series of events. Having very recently undergone a painful break-up from his girlfriend, one afternoon Kevin started crying. He had brought a load of firewood back home.

> I left the rear hatch of the Jeep open, because I intended on going back out to finish unloading. I lay down on the couch with the cat and just kept crying. I am not sure how much time passed, but Wayne W. knocked on my door, screaming my name. He is a friend from Oneonta, who had heard on the scanner that a sheriff had driven up my road, found my Jeep, and seen what he thought was a bear in the backyard. Now, that same sheriff, two state troopers, and the fire marshall came flying up in the snow. I went outside, with Wayne, and the sheriff said that he saw a bear trying to get in my Jeep when he came around the bend down by my property. He said it went into the creek, then when he stopped

behind my Jeep, he could hear it breaking branches up that steep mountain to my west.

Finally, everybody left except me and Wayne. When we went to unload the rest of the wood, the truck was empty but for three logs, and the wood was piled up near the burn barrel, out back. Earlier, I had only just begun the process, myself, so the truck had been full. Odd as it sounds, I began to suspect that something was trying to help me, though I definitely didn't want its help. There are many old apple trees on my property. I thought that if I left an offering at the northeast corner, where my land meets state land, whatever it was would leave me alone. So I left a big Tupperware container, with no cover, filled with apples, and nutty power bars, on a platform, about five feet up a spruce tree. Next morning, gone. OK, maybe deer, or some animal. Next day, same thing, four or five apples, nutty bars, chocolate sugar cookies—gone.

At the time, part of me is saying, "What is wrong with you, Paul?" The other part is starting to feel relieved, like everything is going to be OK. Still, I'm afraid to tell anyone, and it could still be animals.

Third day, I had dinner with my friend, then stayed at his house. I didn't get a chance to walk up the mountain to load the bait container. The next morning, 5:00am, I left his house, and drove up to my trailer. The container is sitting on my back stairs, with this frozen, dried-up little bush: lemon mint. Pulled up by the roots.

I unlock the door, go in, and the window by the couch where I'd started sleeping is open, and another three or four dried-up mint branches are lying on the back of the couch. I really lost it at this point. The cat is nowhere to be found. I start to call the sheriff, but what the hell am I going to say?

> The next day, after I'd left another "gift" in the bait container at the corner of my land, more mint leaves, and the whole plants, appeared in the seats of my Jeep, all pushed through the space I'd left the windows open.

The nature of the habituation experiences that people undergo depends heavily upon context; the quality of the interaction seems conditioned most by the posture assumed and the attitude projected by the *Homo sapiens* involved. This is a lesson well learned, in their own research, by Goodall and Fossey, who had to teach themselves an ever-increasing level of patience and good will, which the primates eventually returned in kind. When they erred, they often paid the price, as in this incident recounted in *Gorillas in the Mist.*

> On a slope gorillas always feel more secure when positioned above humans. I never relished climbing up to a group from directly below, but the thickness of the vegetation compelled me to do so. Once . . . just about twenty feet below the gorillas, who could be heard feeding above, I softly vocalized to make my presence known. A number of curious infants and juveniles climbed into trees above to stare intently down at the unaccustomed equipment [my bulky tape recorder] . . . . Just as expected, the [adult male] silverbacks instantly led the females, all hysterically screaming, in a bluff-charge to within ten feet. Because of the intensity of the screams . . . I tried to bend down to adjust the machine's volume, but the slightest movement incited renewed charges from the overwrought animals. Forgetting all about the microphone, I whispered to myself, "I'll never get out of this alive!" Only when the group eventually climbed out of sight was it possible to turn off the recorder.

The so-called "Siege at Honobia," in 2000, presents a darker side to both human and Sasquatch behavior. It all started when residents of a rural Oklahoma property began leaving fresh deer meat in a freezer in their backyard. On January 17 of that year, the Bigfoot Field Researchers Organization Web site received the following urgent message:

> Too many incidents to mention here, please have someone contact us. This is no hoax and my brother is afraid for his family. This creature is getting bolder every time it returns. This thing is huge, walks upright, smells like musky urine, burned hair type odor. He repeatedly comes back in the early morning hours after midnight and harasses them until just before dawn. It has on more than one occasion tried to enter their home. We don't know where to turn. Everyone thinks we are crazy when we mention it. Please, we don't know what to do but I do know that something needs to be done! There are stories we could tell that would make the hair stand on your neck.

"The message went on to explain," writes Matthew Moneymaker, Founder and Director of the BFRO, "that the family was having problems over the past two years with one or more nuisance animals that were prowling around outside the home at night. The animals were stealing deer meat from an outside shed. The situation had escalated when the animals tried to get into the home. At one point the father went outside to confront the animal. He got a good look at one, and took a shot at what he claimed was a Sasquatch running back into the woods.

"We contacted the family after receiving the report," Moneymaker continues. "The man we spoke with first was the brother of the father of the family. He insisted that they were not kidding around. At least one Sasquatch was coming around the homestead almost every night. It was coming onto the porch, messing with a window, wiggling the door knob as if it wanted to get in, and even stealing deer meat out of a freezer that was kept in an open-sided outbuilding. Whatever it was wasn't alone. The family could hear chattering and screaming from the hills when the prowler(s) were near the home.

"The wife was too fearful to remain in the house. She and the kids were relocated temporarily while the men armed themselves with assault rifles and prepared to defend the homestead against the nightly prowlers."

Moneymaker dispatched a regional BFRO Investigator to the scene, and then monitored events by phone. Here is his account:

> Tim [who shot at the intruder] wants us to take care of his problem. He doesn't want the Sasquatch coming back to his house anymore and he doesn't care what it takes to make it stop. He will not move and not hold back from shooting at it if it returns.
>
> [The BFRO Investigator and two other men] are at the location setting up. I spoke with Tim's wife briefly. She reiterated how frightened they were of this thing and described some of the incidents. Far from jumping to conclusions, she said she and her husband had denied the whole thing to themselves for a few years. It wasn't until after the deer meat (three complete quartered deer) had all disappeared from the large, chest-high freezer in the outdoor shed that the intruder started trying to get in the house at night. It didn't just scratch at the window, she said. It had pulled off parts of the window and was getting bolder in its attempts to get in the house. The recent deer kill found outside had not been shot. One of its legs was violently twisted and broken. It had clearly been carried, not dragged, to the spot where it was found.

The most interesting thing was how the predator pulled out the internal organs. The belly of the deer had not been opened. The opening was up between the neck and rib cage. The predator made a hole large enough to stick its arm in and apparently reached down from above the rib cage and pulled out the organs.

The loud vocalizations, tree thrashing, chattering and whistling outside the house at night are the most noticeable, recurring features. There was considerably more noise during the night the deer was killed.

[Twelve hours later, more details:]

I was on the phone with the people at the house last night for a few hours. I was asking questions and listening to what was happening. Things got very hectic at one point. These guys were actually shooting from the porch while I was on the phone.

From my conversations yesterday and early this morning with the residents, we think we figured out why this situation is so extreme. The underlying cause seems to be that lots and lots of deer congregate on Tim's property. He's got thirty acres in the mountains and he plants Austrian snow peas all over the property, especially near the house, because deer go crazy for them this time of year. People plant these plants specifically to attract deer. It makes it easier to hunt.

Many deer come to feed on his property. There's a deer overpopulation problem in the area to start with, so his property is apparently an effective magnet for deer. There are so many deer that he doesn't even have to get off his porch to go "hunting." He bags lots of deer on his own property and has been doing it for a few years now. He said that on some occasions the deer carcasses were snatched away by something.

A few times Tim ran out after it, but it would always flee into the woods. The first time he got a good look at it was the night he shot at it—a few nights ago. The most baffling thing for all of us was why these things weren't running away after being shot at. They'd pull back a bit in the trees, then move to a different part of the hillside and could be seen through the brush when the spotlights reflected off their eyes.

I asked Tim if he ever spotlights deer at night from his porch. He does. Then we established that indeed, MOST of the time when he's spotlighting the woods and shooting from his porch is when he's shooting at deer, not Sasquatch. So if the animals who aren't running away from the loud gunshots are some kind of predator that's been in the area for a while, then those predators may have noticed that sometimes after those spotlighting-gunshot incidents, a wounded deer would be struggling up the hill trying to get away . . . and will be much easier to catch.

Deer will always take off running when they hear gunshots, especially within fifty yards, that's how they know they weren't seeing deer's eyes while the shooting was going on.

Tim sounded stunned when I explained the deer connection. He slurred out a long steady " . . . oh my God," as if it finally all made sense to him. The Sasquatch might be hanging around the property waiting to grab a wounded deer. I explained that these predators might not understand that they are the intended targets now, because all they would see is a spotlight shining through the trees toward them, then a very loud BANG from an assault rifle. The animals may be expecting to see wounded deer running toward them up the hill. They may have watched that pattern for years.

> It's possible they either don't realize that there are bullets whizzing by them, or they've gotten used to it. At that range the shot is so loud you wouldn't hear a bullet hitting the trees next to you. And they wouldn't see when the guns are pointing right at them because the spotlights would be in their eyes at that moment. It may appear to be business as usual with all the shooting going on.

Eventually, the situation de-escalated when the residents simply stopped killing deer; over time, the visitations waned and then ceased altogether.

A polar opposite approach is advocated and practiced by Robert W. Morgan, Sasquatch researcher for more than half a century, and is pursued as well, each in her or his own way, by all the researchers profiled in this book. "I had my first encounter in 1957," Morgan recalls.

> I had just gotten off a cruise in the Pacific with the Navy. I headed for the mountains. I was in Mason County, WA. I heard something coming down in the brush behind me. It was rustling around, and as I moved to one side, I started seeing black patches of hair, and naturally thought it was a bear. So I stepped out and I yelled, and everything went dead silent. I yelled again, and it started running at an angle, and it got up the slope to me. And finally it got to a point and turned around, and I saw it from just above the navel, up. And I'm looking into the eyes of a Bigfoot. Being a kid from Ohio, I'd never heard of this, so to me I was looking at a gorilla, but it was the most expressive, human-like gorilla. His face looked much more like a man's than a gorilla's, but he was real hairy all over, so that was all I could think. The look on his face was almost comical, because he was as surprised as I was.

Morgan's *The Bigfoot Pocket Field Manual* lays out "sincere counsel" for achieving "passive contact" with this species.

> Never carry a firearm, never even raise a camera to your face, because "most of them have observed this same behavior in hunters." Once you have chosen a research site, it's time to create a provocative routine. Your objective is to be non-threatening enough to be tolerated, yet so different from the scads of usual hikers as to warrant investigation. Cultivate that difference in your own way. Be creative, but never be loud, intrusive, disruptive, or flamboyant. Make yourself and your routines familiar to the Forest Giants. Find a way to gently announce your presence even while hiking. Try whistling or singing a tune now and then. Walk casually enough to allow your own observations but don't bother trying to sneak around because your actions will remind them of hunters. The reason to do all this is to deliberately come under their scrutiny. The most difficult aspect of this passive research method is the interminable waiting: It may take months, even years, of repeated visits to your research site in order to produce results. To be a gracious host to a Forest Giant, you must first design the party with careful attention to detail. Consider your invited guests' requirements: only a friendly and mutually curious atmosphere, the absence of loud noises, quick motions, and sudden lights. Your invitation begins the moment you drive down your first tent peg. However, do not imagine that you will be present the first few times a Forest Giant might drop by. While the Giants are nosey characters, they are considerate yet ghostly visitors. They can appear and vanish in a heartbeat. Unless you are sharp, you may not know they have visited in your

absence unless you set your stage with precision. Your observation camp must be neat and tidy to an extreme approaching anal retentiveness. Prepare to set out some passive lures that will give you a perceptible sign if they are "inspected." Your lures should be too subtle to be noticed even by [human] experts because they will consist of items that are commonly found in camps. Instead of being cleverly hidden to snap or snare you guest, your harmless lures will be even more cleverly place *in the open*, yet the slightest touch will be easily detected. For example, I routinely place on a stump or a flat rock my steel signal mirror, an open plastic case containing aromatic soap, toothbrush, toothpaste, dental floss, and my razor, each arranged with its tip touching a curved line that I faintly etch onto the rock or the stone with the point of my knife. I sometimes leave a book or a magazine, anchored by stones, open to a page that might catch a Giant's eye. I use *National Geographic* or similar publications that contain images of gorillas, chimpanzees, or orangutans in gentle contact with humans. I prefer photos of Jane Goodall or the kindly folks at The Gorilla Foundation as they work with Koko. If you play recorded music, be prudent: make it soft enough that no human being outside of the immediate perimeter can hear it; you must not attract hikers. Also, never succumb to playing popular music because that will place you in the same category as the usual campers. Avoid that comparison at all costs. Why are you doing this? Because that music will serve as your trademark with them, and when they hear you play it, they'll know who you are. Remember, you are involved in a protracted, meticulous chess game in which nobody ever scores a check mate and no pieces are lost.

As I will describe, I myself have taken the opportunity to apply some of this advice in my Vermont Ravine Project, and have reaped exciting, preliminary results.

We could continue to survey, in summary fashion, scores of further cases and styles of habituation, each with its own peculiar context-related twists; but instead, this book will go into depth about just a few.

First, though, we'll take a look at the work of the research group that has managed to pull this creature out of the realm of popular myth and into the light of consistent, empirical study, learning how—throughout North America—to communicate with it in a preliminary, very simple manner, and even, to some extent, how to predict its behavior. Attending expeditions with the Bigfoot Field Researchers Organization became my avenue to learning the basics of Sasquatch behavior, and discovering the existence, and the nature, of habituation sites.

The stage has been set, finally, for a deeper-level interaction, and mutual education, between our two primate species—*Homo sapiens* and Sasquatch—in the tradition of the decades-long, exquisitely detailed projects of Jane Goodall and Dian Fossey.

> Peanuts . . . was feeding about fifteen feet away when he suddenly stopped and turned to stare directly at me. Spellbound, I returned his gaze—a gaze that seemed to combine elements of inquiry and of acceptance. Peanuts ended this unforgettable moment by sighing deeply, and slowly resumed feeding.
>
> Two years after our exchange of glances, he became the first gorilla ever to touch me . . . (*Gorillas in the Mist*)

Titusville Morning Herald

Titusville, Pennsylvania, November 10, 1870

The Wild Men of California

A correspondent of the *Antioch Ledger*, writing from Grayson, CA, under date of October 16, says: "I saw in your paper; a short time since, an item concerning the 'gorilla' whichh is said to have been seen in Crow Canon and shortly after in the mountains at Orestimba Creek. You sneered at the idea of there being any such 'critters' in these hills, and were I not better informed I should sneer too. I positively assure you that this gorilla, or wild man as you choose to call it, is no myth. I know that it exists, and that there are at least two of them, having seen them both at once not a year ago.

"Their existence has been reported at times for the past twenty years. Last Fall I was hunting in the mountains about twenty miles south of here, and camped five or six days in one place, as I have done every season for the past fifteen years. Several times I returned to camp, after a hunt, and saw that the ashes and charred sticks from the fire-place had been scattered about. An old hunter notices such things, and very soon gets curious to know the cause. I saw no track near the camp, as the hard ground, covered with dry leaves, would show none. So I started on a circle around the place, and

300 yards off, in damp sand, I struck the track of a man's feet, as I supposed—bare, and of immense size. Now I was curious, sure, and resolved to lay for the bare-footed visitor. I accordingly took a position on a hill side, about sixty or seventy feet from the fire, and securely hid in the brush. I waited and watched.

"Two hours or more I sat there. The fire-place was on my right, and the spot where I saw the track was on my left, hid by bushes. It was in this direction that my attention was mostly directed. Suddenly I was startled by a shrill whistle, such as boys produce with two fingers under their tongue, and turning quickly I ejaculated, 'Good God!' as I saw the object of my solicitude, standing beside my fire, erect and looking suspiciously around. It was in the image of man, but, it could not have been human. I was never so benumbed with astonishment before. The creature, whatever it was, stood full five feet high, and disproportionately broad and square at the shoulders, with arms of great length. The legs were very short, and the body long. The head was small compared with the rest of the creature, and appeared to be set upon his shoulders without a neck.

"The whole was covered with dark brown and cinnamon-colored hair, quite long in some parts, that on thehead standing in a shock and growing close down to the eyes. As I looked, he threw his head back and whistled again, and then stopped and grasped a stick from the fire. This he swung round and round, until the fire on the end had gone out, when he repeated the manoeuvre. I was dumb, almost, and could only look.

"Fifteen minutes I sat and watched him, as he whistled and scattered my fire about. I could easily have put a bullet through his head, but why should I kill him? Having amused himself, apparently all he desired, with my fire, he started to go, and, having gone a short distance, he returned, and was joined by another—a female, unmistakably—when they both turned and walked past me, within twenty yards of where I sat, and disappeared in the brush. I could not have had a better for observing them, as they were unconscious of my presence. I have heard this story many times since then, and it has often raised an incredulous smile; but I have met one person who has seen the mysterious creatures, and a dozen who have come across the tracks at various places between here and Pacheco Pass."

CHAPTER ONE

Listening At The Tree Line: The Study of Sasquatch Comes of Age

> There's this American chauvinism, that we're well in control of our own country, thank you very much. That *nothing* like this could get away with it . . . in our *own backyard.*
>
> —Matthew Moneymaker, Founder and Director, Bigfoot Field Researchers Organization

BFRO Expeditions 1—What It's Like to Go

It's elemental. Hard, cold ground. Distant outline of hills, of trees. Wind through grass. Breathing humans nearby. Ferns. Silence that pulses, then gradually inflects itself with insect and animal signals. You listen. You try to remember when you've listened more emphatically. Probably not since childhood, since that closet.

Nor have you been awake this late, not happily anyway, not without chewing over anxieties in bed or soothing a frantic baby.

It is 2:16 AM.

During a brief, hushed walkie-talkie exchange between your small group and Matt, the expedition leader, you dare to crunch, gingerly, your last handful of peanut M & Ms, and then it's back to the task at hand. A task that you still can't believe you're actually undertaking.

In certain moods—and your moods shift minute to minute—you think you might as well be awaiting the Mother Ship.

Sure, a few people have come equipped with night-vision goggles, global-positioning systems, thermal-imaging devices, but not you. No, you prefer to go with your very own low-tech ears. You shut your eyes and listen harder. You're sitting hunched and cross-legged in the chill on a tarp at the edge of a cold-dewy field. You're in a bowl-shaped formation, surrounded by those hills, their ridgelines. No moon. Starlight sufficient to show only dim contours of other group members, breathing.

Matt has divided the expedition members into five groups of three or four, and now he radios that it's your group's turn to make a series of "wood knocks" against the nearest tree. You stand stiffly and take up, of all things, a baseball bat. The slam against the oak terribly stings your hands. You send a series of three—BAM! BAM! BAM!—out into the night. Like a message in a bottle thrown into a vast sea.

And then, you resume your seat.

You recognize the luffing call of a barred owl; the chorus, from far up on the ridgeline, of a pack of coyotes; bullfrogs; insects; small animals crackling leaves or twigs in nearby foliage. You have become expert already in factoring out such sounds. What you are waiting for, the whole reason you have traveled across several states to be here tonight, much to the bewilderment of family and friends, is the unthinkable. A reply-knock from deep inside the forest. It may come a moment from now. It may take two more hours. It may never come at all. What if it never comes at all and the joke is on you?

Ten minutes. Nothing. Some kind of bird off to the north. Five more. Matt asks the next group down the tree line, a hundred yards away from you, to knock. You can hear their bat, their oak, though you hope your radio will crackle suddenly with the news, "That wasn't us!"

The idea is to do what *they* do. They vocalize. They make loud "whoops!" and screams. But more often, to communicate amongst themselves in the dark, they knock. They bang thick sticks or logs against tree trunks. We want to sound just like them, to lure them, to *appeal* to them from the other side of this timeless gulf. Because this is how they seem to locate one another across miles in the night. Matt has told us that he and countless expedition-goers have heard such distinctive wood knocks throughout North America. "They make them in Florida, in West Virginia, Ohio, New Mexico, California, Oregon, British Columbia. So since they're not getting together to *plan* this behavior, this must be behavior that's *ancient*."

He is talking about Sasquatch, a species of enormous primate that has, apparently, been able to elude our capture and domination—even to elude human *belief*—for eons. He is talking about none other than the animal that has captured *your* imagination since, at age ten, you first saw the Patterson/Gimlin Film.

One reason, you have figured, that many find it difficult or impossible to believe that Sasquatch can be a real creature surviving in our forests is that they radically underestimate the nocturnal world. It is at night that the animal is most active, and especially between the hours of midnight and dawn, when there are approximately 1% as many people out of doors as during the other eighteen hours of the day. And of this 1%, how many of us stray off of roads, yards, public spaces? Well, you finally have found an organization that does exactly that.

It is 2:37 when, sure enough—and never has this common phrase, *sure enough*, seemed so ridiculous, or so right—you hear, if not as clear as day then half as clear, maybe a hundred yards off inside the forest, a responding series of knocks: two, a thirty-second pause, and then two more. After everyone takes a couple beats for absorption, all five groups quickly call in, reporting that yes, they have heard the knocks too, and that no, none of them is playing a trick on the rest. This hardly needs asserting, since the knocks came—you need to keep and keep reminding yourself—from *inside* that forest, away up the side of the hill toward the ridge, and all people are accounted for here at the edge.

As per instructions, then, everyone falls into radio silence. "If you hear wood knocks, that means he's curious," you've been told. "Just keep quiet and give him time to come closer."

After some six minutes, the longest six minutes you can remember, from another ridge beyond and above the opposite side of the bowl-shaped field comes a responding knock, just a single one this time, and softer, farther.

Nine more minutes pass, and the next sound you hear is a heavy footfall through underbrush, approaching down the long incline. You warn yourself not to jump to conclusions, that it could be a deer, a moose, a bear, but the steps are unmistakably bipedal, a *stamping* that makes no effort to conceal its power. Quite the contrary. CRACK! A hefty tree branch is split, and splits the night, and you abruptly recall others' reports of hearing such displays in the woods. Intimidation. Reprimand for territorial encroachment. And . . . it works. You back away from the forest like you're playing crab soccer. You feel like you might faint or have a coronary. Mike, to your left, breathes, "Hoe. Lee. Fuck."

But your equilibrium is restored by Matt's calm voice over the walkie-talkie. "Okay, folks, this is a *good* thing. Don't be afraid. This is normal behavior." He has been doing this for twenty years, and never gotten closer than forty feet; this underscores the idea of an inviolable barrier, maintained by our enormous cousins. "Whoever thinks they're closest, go ahead and try to talk to it. Show it something it's probably never heard before from people, which is a little sympathy and understanding. Remember, it's never known humans to act like we're acting. It is not going to hurt us."

Your mind boggles—you can literally *feel* it in the *processes* of boggling—as you ask it to accept the premise that a seven- or eight-foot-tall, nine-hundred-pound *primeval ape creature* is actually standing in the woods over there. Inhaling and exhaling the neighboring air.

All you want to do is to train a piercing spotlight in there through the trees. But you don't have one, nobody does. They're strictly prohibited on BFRO expeditions, because bright lights are adversarial, that's what hunters do, or mere campers. "If we don't make any aggressive moves," Matt has taught everyone, "then the animal will not quite know how to categorize us and will be curious. It'll be more likely to hang around."

Jonathan, in the group a hundred yards to your south, radios that he'll give talking a shot. He has been on several expeditions; you even shared franks and beans with him earlier tonight, at his campfire, and he related past encounters. He seemed so happy back then, poor doomed fella. You can't hear his exact words from this distance, but his tone is very friendly, veryveryvery friendly.

(The next morning, at late breakfast, he'll tell you, rather sheepishly, what he said. By the light of day, the whole incident seems, of course, positively absurd, dreamlike, instead of like any dire showdown. "'Hello,'" I said. "'Hi there, sir. How are you tonight? We're just here to learn about you. We respect you.' At one point, I think I actually might have said, 'We come in peace.'")

Jonathan receives no response. Minutes pass, and intermittently he tries again. And then, just when you're ready to assume that whatever it was has stealthily departed, or, more accurately, that it was never there to begin with, was instead the result of some sleep-deprived collective wish-fulfillment hysteria, comes the belated reply in the form of the leaf-ripping flight of some object, and then a thud onto the sod, not fifteen feet from Jonathan's group. A rock has been pitched a long way through the forest, and it's the size of a basketball. (Barbara finds it on the way back to base camp, and you all claim it as your prize, put it front and center in your group photo.)

You hear footsteps retreating up the hill for half a minute—this time just stepping lightly, crunching in dead leaves—and then nothing more, so the small groups reconvene and head back to base camp and your dry sleeping bags, exhausted, freaked out, and deeply honored.

Later tonight—and yes there is room for a "later" squeezed in before dawn—one couple in their tent wakes at 4:34 AM to approaching footsteps, but this time they're entirely different, not aggressive and heavy but sneaking near the vinyl wall, stealthy, delicate, though clearly bipedal. And then something grabs and shakes the rain fly for twenty or thirty seconds.

In the morning, over breakfast, Matt calms the couple by explaining that what they experienced is normal behavior. Often, first, the Sasquatch attempts to intimidate the humans, to oust them from territory where, long after dark, they don't belong. Second, once the people have gone deeply to sleep, and are blatantly snoring, incapacitated by slumber, between, say, 3:00 and 5:00 AM, the Sasquatch's native curiosity can afford to supplant its fear and affront, and so it can investigate.

Other middle-of-the-night stake-outs follow, and you experience nothing. Nobody does. Nothing at all. The nights ring back at you with hollow silence. The quarry has either moved on or decided it is unsafe to play.

The expedition members, though, draw together and become close. You find in one another kindred spirits, people who have endured years of being the butt of jokes but possessing, nonetheless, the necessary wit and imagination to grasp the very possibility of Sasquatch, to take in and take seriously the extensive physical and anecdotal evidence. How can you *not* appreciate such folks, so bold and wide awake? And what an ancient, potent archetype you're co- and re-creating together, too: humans versus "the monster."

Over the next several expeditions, you learn that in fact the majority of the time group overtures go unanswered, even the great majority. Because in order for acoustical contact to occur, the animal or animals must be within earshot *and* be in a mood to communicate. After all, they are in their element, behind trees at the pitch-black core of night. You come to think of it this way: here is a species whose survival strategy, vis a vis *Homo sapiens*, has been virtually 100% effective for hundreds of thousands of years. And even when they do sound off, occasionally, they are simply not going to come close, not going to commit that kind of fatal error. If they did, we'd all know about it; the case would have been closed long, long ago. There are researchers who have spent their entire adult lives in this pursuit, such as Grover Krantz, John Green, Rene Dahinden, and never caught a glimpse. And here you expect to witness a breakthrough equipped with your $250 SONY "nightshot" Camcorder and a fistful of peanut M & Ms?

But two months later, in May 2006, on the Ohio Expedition, in the early afternoon, Florida-based legal secretary Caroline experiences a daylight sighting, rare beyond rare, while standing on a defunct railroad bed, looking up to a ridgeline.

> At the top of the hill, you could see blue sky, and then you could see all the saplings, and you could see between the saplings and the blue sky a figure walking. I thought it was a person, but then we told Patty to go in the same place and I could see the colors. I could see the color of her face, I could see the color of her clothes. And this was just all one color—torso up, dark . . . Matt and Patty, when they were standing up there, you could definitely see the difference between where their head stopped and their shoulders began. This thing had shoulders, but it wasn't a definite neck.

Twelve hours later, at 1:41 AM, you produce an impressively sustained moaning howl (emulating the sound recorded in 1994 in this same county) and receive a response howl from the ridge high above. It is, to put it mildly, a kick, and you entertain the fleeting fantasy that you could now simply charge up that slope and meet the source of that glorious call. But then of course you realize such a climb would take you two hours, by which point the caller him- or herself, having recognized the first few *moments* of your heroic ascent, would have exited to the neighboring county.

Nearby, the same night, Onil (a Toronto S.W.A.T. team member) and three other expedition-goers walk along a slow-flowing river, at a place called "Gretchen's Lock," named after a small girl who died of malaria and was buried here back in 1838. Her ghost is said to have haunted this spot, making her presence known these 168 years since. Out of nowhere, Onil's group hears the great plunging impact of a rock thrown into the water less than ten feet from where they are standing. Judging by the sound, this object was about the size of a volleyball, and based on the direction it must have come from, was heaved more than eighty feet, across the wide river.

Such incidents, as well as those of stick- and log-throwing, are commonly reported in areas with a long history of Sasquatch encounters. With remarkable consistency, these projectiles land close to people but do not strike them; it seems a good way to spook us, to chase us off Sasquatch turf. And it generally works wonders. Some of the locals, whom you talk to, seem to prefer a different explanation: little Gretchen's ghost has always been credited with a particular mischievous penchant for hurling rocks at interlopers.

The next morning, on a trail system above Gretchen's Lock, you're in a party of four who comes upon an extremely impressive "tree structure" in the forest. It's made of seven trees, pushed over from where they grew, averaging ten to twelve inches in diameter. These trees intersect in a neat cross-hatch fourteen feet off the ground, leaning into the crook of a larger, upright tree. Choosing a tree of the same thickness, growing nearby, burly, six-foot-five-inch Mike from New Jersey finds that he is unable even to budge it, much less to topple it over and add it to the formation.

Throughout North America, such stick and tree structures—most much smaller and less elaborate than the one you found in Ohio—have appeared in the vicinity of reported Sasquatch encounters and acoustical evidence. Researchers speculate that they serve as some sort of marker, rather than as shelter, but it's not possible at present to be more precise. Given that they are often found near roads and trails—and often thirty to forty feet off of such routes—one possibility is that they are built there by the older ones to warn the younger ones to avoid human areas. They may also represent aesthetic expression, or play, or the simple statement: *We were here.*

On the Arizona Expedition, while exploring along the steep wall of a rugged box canyon within the White Mountain Apache Reservation and just two miles from where Apache rangers found massive five-toed tracks in the mud, you and three other members discover a horizontal cave thirty-five feet deep, shaped like a keyhole. Its first fifteen feet, the narrow stem of the keyhole, is an entrance corridor in which large rocks have been cleared away, moved to either side, leaving a smooth floor for crawling. Bears, according to Lynn, a wildlife biologist attending the expedition, have never been known to do this in their dens; instead, they simply clamber over such rocks.

This corridor then opens out into a round room twenty feet in diameter. Back here, the floor is covered with a uniform eleven-inch-thick layer of soft organic material, mostly dead grass and leaves (deeper, this has decayed and disintegrated into dust), that has been carried in here and spread out as though for bedding. Lynn can think of no animal capable of dragging hundreds of pounds of soft organic material far into a cave, no animal without *hands*, that is.

Nowhere in the whole place is there any sign of human use—no fire ring or soot on walls or ceiling, no beer cans, cigarette butts or a single scrap of litter. The rounded chamber at the back of the cave could easily sleep a family of three. And it would be temperature regulated: in winter, the body heat would make it nice and toasty, and in the summer, one could duck in here to beat the heat.

Furthermore, this cave is located strategically perfectly: thirty feet beneath the rim of the canyon, its mouth entirely hidden from up top, and overlooking, three hundred feet below, the confluence of two rivers where the elk cross, and come to drink.

It's actually quite homey inside here and, reveling in your discovery, you lie down, curl up, pretend to be many times your own size, in which case the walls would curve neatly against your spine.

I can now shift from second to first person and tell you that all of the above did occur to me personally, or to credible others around me, during eleven BFRO Expeditions between September 2005 and October 2008. (Much of what I have recounted is documented on *Impossible Visits: The DVD*.) Though moments of close approach are few and far between, the mere fact that these expeditions *ever* yield genuine contact at all, even from a distance, is remarkable enough, representing as it does a quantum leap over previous efforts, especially during the second half of the Twentieth Century, when "expeditions" were little more than hunting or scouting trips, poking around in promising areas during the day. Before the BFRO began finding success in the mid-1990s, researchers did not employ the subtle approach, making consistent nocturnal overtures, attempting to elicit response.

For example, I once drove eleven hundred miles round-trip from Vermont to an expedition in southern Pennsylvania, stood in the woods with fellow members at multiple propitious locations in the middle of the night, making wood knocks and whoops, attempting to emanate a maximally peaceable attitude into the pitch blackness, and heard nothing back from across the divide for the entire three days. Yet I reminded myself that this outcome is, after all, part of the *point.* Theirs is an entirely strategic lifestyle, and it is, of course, thanks to this very fact, this foundational cold shoulder, that Sasquatch has been able to persist into our own time. If this creature had ever been attainable in the first place, all we'd ever hope to attain today would be its dry fossils.

BFRO Expeditions 2—Fostering Mini-Habituations

Unlike the case of chimpanzee and gorilla research, the field of Sasquatch study affords, as of yet, no luxury of protracted witness and interaction; thus, it remains in its infancy—that is, the field *as a whole* does. Many avid and committed individuals have, however, interacted with Sasquatch *in* the field, at certain forest locations or in their own backyards, but have kept these interactions strictly *entre nous.* I learned this important truth only gradually. Even at habituation sites, any interaction with this species is a glancing one, oblique and capricious, though developing a relationship with an individual or a family, over months and years, is the surest way of shortening your odds.

In terms of evidence officially documented and reported, the best that had been achieved before the mid-1990s was the chance wilderness or roadway encounter, in which both human and Sasquatch are startled and so neither party is acting naturally. And thus, the person almost never has the presence of mind to wield a camera. (Notable exceptions are Roger Patterson, 1967; Paul Freeman, 1994; and Lorrie Pate, 1996.)

So Matt Moneymaker spends a lot of time during each expedition educating participants on what to expect, rehearsing outreach strategy, going over dramatic encounters from expeditions past—none of which culminated in a proximity closer than fifty feet, much less a face-to-face encounter—all in an effort to help quell that primal interior voice that tells us we're dealing, here, with a monster. He frames the issue this way: "It's as if you're trying to make contact with a lost tribe. You approach them with great respect, with the attitude that you know they're smart, and that you want to let them *know* that you know. And that you understand you are trespassing on *their* territory."

He encourages us to clear aside the confusing clamor of apprehension, to quietly explore and tune in to our intuition, to become alert to the subtle sensations that may arise. Sometimes, one can actually feel something in the air, a shift in quality, becoming aware of being watched or receiving what he describes as "a *sizzle.* You know, the sense that I've been here before . . . but I haven't?"

Expeditions last three days, occasionally longer, and there is a reason for this. "If they're around they can obviously be very cautious and very coy at first, and you won't hear much. And they'll come around and check you out and you won't even know they did it. But after a couple of days, they may just begin to let down their guard. One night, and you're showing good behavior, and they're still playing it cautious. Then the next night you're still making their sounds, and you're not grabbing spotlights, and you're not pointing guns and going out after them every time you hear something. After a few days, they may begin to let down their guard a little and let you hear them, and even see them, usually at night."

In North Carolina, once, I was sitting right next to Michael from Pennsylvania when he spotted, through his thermal vision scope, the shape of a massive figure, like a tall football player with shoulder pads, walking among the trees at about seventy feet. He turned his upper body toward me and the others, said, "Our friend is here," then took a moment to switch on the record function (limited battery life prevents him from running recording constantly). When he turned back toward the trees, the figure had already receded from view. Michael was despondent. A veteran North Carolina tracker commented, "Them buggers is slick." The thermal scope resembles a gun.

That same night, our youngest North Carolina Expedition member, the fifteen-year-old daughter of an attending couple, had the traumatic experience of being "zapped." She was just sitting at her family's campsite, near the campfire, fifteen feet from the edge of the woods, pretty relaxed, when suddenly she was overcome with anxiety, felt nauseated and weak, with an intense pins-and-needles sensation in her arms. Her parents helped her walk over to join the rest of us, to help give her perspective on what had just happened to her. It was a warm night but she felt profoundly cold; I brought her a blanket from my car.

Certain species of mammals, such as whales, elephants and lions, can emit "infrasound blasts," pulses of sound at a frequency too low to be heard by the human ear. They will use infrasound to communicate over long distances, or to overwhelm their prey; the pulse affects the central nervous system. Lions have been documented disabling gazelle in this manner. There is abundant anecdotal evidence that Sasquatch, too, sends out infrasound blasts, and researchers often refer to receiving such a pulse as being "zapped." Perhaps, in this instance, the Sasquatch decided we had been encroaching on its territory (walking in the dark late at night outside our campground, where human beings are not *supposed* to venture), and so targeted our youngest, and most vulnerable, member.

People can become so profoundly disoriented that they "freeze up." In 1995, I met a man in British Columbia who told me that when an adult male Sasquatch crossed the road in front of his car, then looked in at him, "It pretty much lock-jawed my whole body." While thus paralyzed, victims of infrasound often find that their perception of time is distorted, such that minutes can feel like seconds.

Another secret of the BFRO's unprecedented success rate lies in site selection, both in terms of specific region and then in terms of targeting within that region. There is a careful narrowing process that often takes the first two of an expedition's three days, driven by a particular constellation of topological/habitat traits. I can't reveal these here. Why? Who might misuse the information? Matt answers with one word: "Shooters."

"What we may discover in this whole thing," he conjectures, "is that there's always been this kind of unquenched thirst on their part to interact with people and they just couldn't. Because you know how people react when they see these things. We're not friendly.

"They've seen human activity that seemed endearing or sympathetic. You've probably watched people having fun somewhere, from a distance. And you wish you could go and have fun with them too. I think they watch us and probably plenty of times they wish they could just freely walk among us. They know they can't."

The way to habituate Sasquatch, even during our brief visits to a given site, is to play upon this desire for interaction, and to attempt to walk a tightrope—both emotionally and tactically—between over-eagerness and timidity. If we get a knock response, for instance, we will often refrain from following up immediately with another overture. If they're knocking, they're curious. They may not be sure of the source of the sounds, whether human or their own kind. Best to let this uncertainly ride for a while, see if they will approach for a closer look and listen.

In southern Pennsylvania, Matt and Steve Willis (U.S. Army, Retired) were standing by the woods after midnight, making occasional vocal overtures. "Before they actually came up on us," Willis reports, "Matt said he thought he could hear a faint response to his whoop call. We stood there for five, maybe eight, minutes, and we could hear something coming through the woods, coming toward us. So we just stayed quiet and it kept getting closer and closer. It was just like this, the dark of night, we didn't have any night vision with us, and all we were doing was just listening. At one point, Matt whispered to me, 'If these are 'squatches, we'll hear one go around behind us.' Sure enough, pretty soon that's what happened. The steps started separating into two distinct sets, two pathways that they were taking. There was one group that went directly across the front of us, and it probably wasn't more than fifty to sixty feet away from us. It was an adult and a young one, you could tell, because every once in a while the trailing one would have to run a couple steps to catch up."

"See, you want to have the attitude," Matt tells us, "that you're so sure they're civilized and decent that you can fully surrender yourself, put yourself out there. And that's going to be expressed in all kinds of ways. There's the way to express it mentally, if there's anything to that, and again, you've got nothing to lose by trying. It projects also through your body language, and then just everything that they can sense about . . . is this guy afraid? Have I got this guy scared?"

While I traveled around the country with the organization, eventually rolling up a total of nineteen thousand miles, I began to pick up a few details from other expedition members about an astonishing situation currently unfolding in Kentucky.

It seemed that a woman had succeeded in establishing a reliable routine of feeding a young female Sasquatch. This woman had contacted the BFRO and, in the early summer of 2005, Matt had flown down there to investigate, and her story checked out. He managed to obtain some late-night infrared video footage of the creature coming for a plate of pancakes, eating them with its back to the camera.

Naturally, upon first learning of this Kentucky situation, I became fascinated by this whole new notion of a habituation site. That such a thing existed at all was quite a revelation to me, transposing the spotty tune of Sasquatch research into a rich new key and tempo, and I set out to learn as much as possible.

In the spring of 2005, a young, female Sasquatch began making repeat visits to a particular rural residence in Kentucky. She'd be enticed there by the woman of the house, who'd spotted the animal several times from a distance and had then begun calling out to it in a high, sing-song voice: "Come here, my little man, Mama's got something good for you! Come here, my little man!" (Later, the woman saw the creature squat to pee, and understood that it was female.) The Sasquatch was not very tall, no taller than the woman herself. It was hair-covered and bore a large, rounded head with lots of frizzy, flowing hair running down beside the face. It was barrel chested but immature, with no obvious secondary sexual characteristics.

During the day, she had no luck in getting this bizarre primate to approach. But late at night, she'd sing her little song and leave a paper plate of pancakes and syrup on a low dirt hill, near the forest, at the edge of her backyard catfish pond. Remarkably enough, twenty minutes or half an hour later, the individual would consistently emerge from the densely wooded valley and accept the offering. Sissy could hear it down there, and in the morning, she'd find its footprints, impressed into the dirt. These prints went far deeper than her own, or even those of her husband, at more than two hundred pounds, confirming how robust the visitor was.

"Of course it makes sense that the subject was a juvenile," says Matthew Moneymaker, "because she hadn't learned the ropes yet. She let down her guard." The organization's Web site, www.bfro.net, has received many thousands of reports of encounters from all over North America, has interviewed the witnesses, assessed their credibility, and concluded that Sasquatch species distribution is much wider than formerly contemplated, even by those who believed it existed. But never before had he stumbled upon a case remotely as promising as this one.

In the late-night, infrared footage that Matt obtained, one can first see the figure of Sissy herself, inching along unsteadily in the inky blackness (humans cannot see by infrared light), setting down the paper plate of pancakes, and then withdrawing. Twenty-one minutes later, another figure enters the frame, reaching a thick furry forearm toward the plate. The hand even scoops up a couple of pancakes that have fallen off, replacing them. (Apparently, these eyes *can* see by infrared.) The subject then picks up the plate and sits down on the dirt hill, turning its back to the camera. This reminds one of a rear view of a mountain gorilla—that same thickset frame, the odd stiff straightness to the back, the air of power in repose.

From the movement of its head and from the elbows projecting out on both sides, one can then infer that the animal is eating. And the head looks strange, definitely *not* resembling a gorilla's; it is huge and round, like a pumpkin haloed by wispy hair. A gorilla's skull bears a conical shape or "sagittal crest," as does the figure seen in the 1967 Patterson/Gimlin Film, a female Sasquatch retreating gracefully from the camera. But juvenile gorillas do not yet sport such a crest. So when it comes to the creature on this Kentucky hill, either we are talking about a similar state of morphological immaturity, or else the Sasquatch species features regional differences, like races or breeds. Indeed, Kentucky lore about wildmen in the forests sometimes referred to them as "Ole Pumpkin Head."

This initial footage is somewhat dim and would never convince any hard-core skeptic. Luckily, though, more and better was to come. Much better. Though I have not yet seen it, I have had the other (daylight, color) video clips described to me in extreme (and matching) detail independently by three people who *have* had this privilege. Due to copyright restrictions, I have just cut several pages from this book in which I tried to convey these descriptions, as well as some of the astonishing goings-on behind the scenes. A documentary presenting all of this material is currently in production and will be released before long.

Suffice to say, for now, that at least one of these clips will become as iconic as the Patterson/Gimlin film.

My research has taken me elsewhere, as I've been lucky enough to develop trusting relationships with five primary figures (and some of their family members) interacting with Sasquatch at five other habituation sites. Some of this research has been ongoing for years. I share these stories in this book in the hope that they will contribute to a general increase in public understanding. If the world can become more familiar with the subtle and civil ways of this species, can come to recognize just how widespread is its distribution throughout our continent, then perhaps the "Sasquatch fever" that will ensue after the Kentucky situation is brought to light, and then even hotter after the grim day when a specimen body is harvested, may become tempered by wisdom.

Nevada State Journal

Reno, Nevada, July 16, 1924

Clue to "Gorilla Men" Found—May Be Lost Race of Giants

"Mountain Devils" discovered at Mount St. Helens, near Kelso, are none other than the Seeahtik Tribe, said Jorg Totagi, Clallam Tribe editor of The Real American, an Indian national weekly publication, in an interview here today.

The Indians of the Northwest have kept the existence of the Seeahtiks a secret. Partly because they know no white man would believe them, and partly because the Northwestern Indian is ashamed of the Seeahtik Tribe, said Totagi.

"The 'mountain devil,' or 'gorillas,' who bombarded the prospectors' shack on Mount St. Helens, according to the description of the miners, are none other than the Seeahtik Tribe, [who were] last heard of by the Clallam Indians about fifteen years ago, and it was believed by the present Indians that they had become extinct. The Tribe made their home in the heart of the wilderness on Vancouver Island and also on the Olympic Range.

They are seven to eight feet tall. They have hairy bodies like the bear. They are great hypnotists, and kill their game by hypnotism. They also have a gift of ventriloquism, throwing their voices at great distances, and can imitate any bird in the Northwest. They have a very keen sense of smell, are great travelers, fleet of foot, and have a pe-culiar sense of humor," Totagi added.

"The Seeahtik Tribe are harmless if left alone. The Clallam Tribe, however, at one time several generations ago, killed a young man of the Seeahtik Tribe and to their ever-lasting sorrow, for they killed off a whole branch of the Clallam Tribe but one, and he was merely left to tell the tale to other Clallams up-Sound.

"Henry Napolean of the Clallam Tribe is the only Indian who was ever invited to the home of the Seeahtik Tribe. It was while Napolean was visiting relatives on the British Columbia coast about thirty years ago, that he met a Seeahtik while hunting. The giant Indian then invited him to their home, which is in the very heart of the wilderness on Vancouver Island. Napoleon claims they live in a large cave. He was treated with every courtesy and told some of their secrets.

"Some Indians claim that during the process of evolution when the Indian was changing from animal to man the Seeahtik did not fully absorb the tamanaweis, or soul-power, and thus be became an anomaly in the process of evolution. It is generally their custom to frighten persons who have displeased them by throwing rocks."

CHAPTER TWO

"Playing Hide 'N' Seek With 'The Monkey Kids'": Five Families Get to Know the Neighbors

The people who are habituating already know what they have living on their land, so really aren't trying to convince others. The satisfaction they receive when they capture the shadow folk in photos is just a bonus for them and just convinces them that what they saw and heard was a real creature. Generally the only telltale sign the shadow folk were there will be foot tracks left in the early morning dew. Occasional sounds or some stick signs. On rare occasions a glimpse of them moving about.

—Anonymous

Family A: North Carolina

The speaker is a forty-eight-year-old woman whom we will call Ammi. Her husband "hates" what has been happening on their property for the past five years, but her son, now twenty-two, has worked with her to learn more. Beyond their yard is extremely thick forest and a vast swampland that extends seventy-two miles northeast. Occasionally, her six-year-old granddaughter visits and talks to "the hairy kids in the woods."

From 2002

It was back when we were living somewhere else, five years ago, and out here remodeling this house every day. Very first thing was a god-awful, gut-wrenching scream that came from our woods one day. My son and I both thought it sounded like a young child being raped or torn apart. There is an elementary school near here, and we believed someone was hurting a child in there. We both ran into that area of the woods, and looked, but saw nothing. I went back to the field, to make sure someone wasn't getting away, and my son continued to look around in that area of the woods. He said he never saw anyone, but it looked like someone had been there, because a lot of plants and grasses were laid down and flattened. We heard the same scream, in the same area, a few days later, but same results, on searching. There was an obvious path we hadn't noticed before, so my son checked the area off and on, at random times for about a week.

Then, from different parts of the woods over different times, we'd hear what we thought was humans trying to break into the place or coming up here. We weren't staying over here all the time, like I said. We thought there were people running through the woods. We shot at them. We've called the Sheriff's Department out here several times, too many times. Nobody every found anything. My son chased them, never could catch up with anybody. And that went on for a long time.

What really brought it to a head was, we rented a backhoe and went to dig the ditches, and when I did that they started throwing stuff at me. Mud and sticks. I really didn't know what was going on, so I had some local investigators come out. They came out three different times and did an overnight. The first time they came out they found footprints and they did some hollers and got some answers, did some wood knocking and got some answers, and that kind of blew us all away. We were like, Oh okay, we got Bigfoots. I mean, we kind of thought the place was haunted for years. But then once we found out, my son and I just really got interested in it. We started trying to study them. Daily. We started going out there and trying to communicate with them, and basically just went from there.

I've got one group on one side of the house that's pretty receptive and friendly and we've come quite a ways. On the other side, behind the barn, they're mean. I've only been doing this about three months.

They hide really good. Even right in front of you. How would I know that? By taking pictures behind my back. I can hear them, and know they come very close, and from the sound of movement when I turn around, I suspected they may show out behind me. So the thought came to do the backwards camera trick. I pretend I am shooting in another direction, and just turn the camera backwards. Then I compare these shots to the ones I take straight on.

I also know by hearing them, all the sounds they make, but hardly ever seeing them. By watching my woods change here daily. In all these years I've only ever seen them three times.

You would just have to experience the thickness of my brush. You could hide an elephant in there. The pictures I'm taking now are winter pictures. In the summertime, those vines, you can't see from your waist down. They could crawl around under your feet and tickle your toes and you wouldn't know it.

I believe what makes them stay here is that our area is very quiet, with a good source of food and water. There's not a lot of traffic. We don't go out in the woods a lot. They've been here before we were here. It's not that they moved in, they just didn't move out.

The nice ones, when I feed them, sweet potatoes and apples and bananas, they sometimes leave me gifts. I keep them all in a small bowl. Pretty rocks. One of the rocks is studded all through with large garnets. One looks like some kind of petrified bone . . . maybe a hip and part of the socket, from some little animal. The rest are lots of quartz, and some is just plain rocks. I also got a civil war grapeshot cannon ball, and half of some kind of old metal bullet mold. And I got a little ceramic duck, that looks like it has spent some time in the swamp.

The mean ones, out behind the barn, they killed my dog and left me his skull and some bones on the work table over there. He must have been bothering them. It was very hard to accept, but we moved on. I have had worse times. You get over stuff, and move on.

I want to let people know that there are some down sides to this too. I will study them while they stay, but if they moved on tomorrow, I would just get back to a "normal" life here. I wouldn't go looking for them anywhere else. Some days I don't want to deal with them here. But I try to keep it steady, so I can learn more.

It's a whole game of building trust with them, and it doesn't just include just handing them food and walking away. People would probably think I was crazy. I go out there and I talk to the woods. I walk around, I speak to them, I sing to them, my son plays guitar for them.

My son is twenty-two. He and I are the only ones doing the study here. He occasionally brings in one friend, and it took repeated visits, but now they come close and vocalize when he is here also. He is eighteen.

I definitely know they know when I'm there. I think they know every time I step out this door because I hear them whistle. They give little whistles to I guess kind of let each other know that you're coming out. From area to area in the yard they whistle the same way. Different little whistles different places.

They imitate my son, and say, "Mom!" loudly, they can sound just like him. I thought he was playing tricks, until they did it when I was here alone, and then when he was with me talking to me

When Rita [another researcher] was visiting here, the one who tries to imitate me followed us from the house, to the swamp side, and kept whistling to us. I've been trying to teach him the tune from "Kill Bill" (when the nurse is walking down the hall). He is really trying to get that tune down, and gets quite persistent with it, if I don't do it back for him. He was back out there again today with me, and trying again. He has the first part down, and is getting pretty good at the second part. He is only about two notes off now, and he whistles the two, but not in the right tune.

It is so amazing to hear it, because it is so different than any other sounds of the woods. At times, he just keeps repeating it over and over, like I do to teach him. They are such mimics! I always wonder if he is the same one who follows me around and does the "babbling brook" noise. Rita has heard that one . . . . and she can explain it to you, because she got such a kick out of it here.

From Rita, an independent Sasquatch researcher, who first visited the property in December, 2007

[transcribed from an interview on the excellent on-line radio show, "Let's Talk Bigfoot," at blogtalkradio.com]

Poor thing. She was taking pictures with a 3.5 MP camera with no zoom through a pair of binoculars. That just tugged at my photographer's heart strings. You know, if she's got Bigfoot in her backyard, she needs a decent camera.

She pays close attention to her swamp and monitors it daily for every little change. When she shares a photograph, she's already compared it to reference photos, so she knows that the figure in the shot is worth a closer look.

I saw this for myself today. Imagine seeing a large form of deep inky blackness behind some brush. So dark that it absorbs all light—like black fur would do. And then suddenly, silently, it's gone and you see browns, tans and greens where it once was. Someday, we'll get a great shot, or a shot that becomes great with the right enhancement. And it may take a long time to get that great shot. Or it may be tomorrow.

I saw more than one dark figure that disappeared after we spotted it; I heard knocking, whistling, bad dog bark mimicry, two incredibly loud and angry thumps, and other odd noises. I had one mimic my whistles. I saw tons of stick structures, trails blocked, trees snapped and in some cases huge logs that had been moved from place to place for some unknown purpose.

It's my opinion that through situations like hers, and with her determination to study these creatures in a peaceful setting, we will learn more about them than we could through other means.

So, we sat around and chatted for a little while, and then we went outside. She feeds them fruit and sweet potatoes. She called them, and I actually heard them walk toward us. There was a group of about six or eight walking toward us and you could hear the leaves just crunch crunch crunch crunch. You couldn't see them. They started walking from a distance of I'd guess fifty yards or so, and approached probably as close as about sixty feet. But the entire time they were behind this screen of evergreen, brush, and vines. They have actually built screens in her swamp made out of these vines and trees.

You could clearly hear them approaching and that was just a fantastic experience. I mean, I was blown away, and I thought I'd be able to get a picture of them because there was a break in this screen, and I thought, Okay, when they walk across that I got 'em, I'm gonna get a picture. But oh no, they stopped just as they got to it. They were just too smart to come out in the open. And we heard some whistling, and we heard the chit-chit-chit-chit-chit of maybe rocks or Ammi was thinking walnuts, being clicked together. And they imitated the sound of a babbling brook. I found that actually kind of humorous because babbling brooks don't start and stop like you flick a light switch. And they also don't move. And during the course of the afternoon, as she was showing me around the place, the babbling brook followed us, which I just thought was funny.

So she basically threw the fruit and the sweet potatoes toward where they were. I think they just see where stuff lands, and then they come later and pick it up when she's not around or when nobody's looking, or possibly wait until dark.

We walked around a little bit and I was trying to get some pictures, and it was really interesting because you could actually see dark fur appear behind the brush, and then it would disappear. And of course she's got several photos that look like that, it's just a dark form that you can see through the brush. And it's challenging from a photographic standpoint because it's hard to photograph anything black anyway, because black is black because it absorbs all light, so if you're talking about dark animals in shadows, you're just out of luck. But that doesn't mean we don't try.

One of the more profound experiences I had that afternoon was getting "zapped," twice.

This was over on the other side of her yard, where they're not quite so friendly. It's where they had put the skull of the dog that they had killed on a work table. The table is strong. You know, it's one of those old-fashioned tables that a farmer would have put together to work on outside, really hefty. We looked at the table and thought, Okay, so this is where the skull is. And then we walked down into the woods there. All of a sudden I felt nervous, I felt a little nauseated, my head was bothering me. I felt very unwelcome. She started some similar symptoms. We stayed down there a few minutes, took a few pictures, and then we came back out. We walked a little ways further down and ended up with some very thick brush between us and the barn so we could not see the barn. And we're poking around, trying to take pictures and stuff, and all of a sudden we hear this THUNK! up at the barn, and we're like, What was that? And then we heard it again. So we figured that we had upset someone that came up and pounded on the table.

I walked up there, with my video function on my camera running, and didn't see anything, that I *remember*, because once I got over there and I turned off my video camera I got another infrasound experience. It's like when you're about to put your hand on a TV screen, you get that tingling all over the surface of your hand, it was like that all over my whole body, and worse on my arms and legs. And I said something to her about it, and then I stepped backwards.

In my mind, it only took two seconds. But she informed me later, Oh no, you were frozen and staring straight ahead for about twenty seconds. And . . . I didn't take her seriously, but I happened to have an audio recorder in my hand, so I played it back and you could tell by the beep when I turned off the video function on my camera, and when I spoke to her. I actually spoke to her twice, and then you hear when I finally do step backwards. It was a total of thirty-six seconds.

It's very interesting to me that I only remember it taking a couple of seconds. So when you hear about some people becoming disoriented and possibly losing time when they have an infrasound experience, I feel like that was an example of it. It was a very profound feeling, I've never felt anything like it. And it was a little scary too, I'll be honest, it was kind of scary. And she didn't go over to that area until a couple of minutes later, and that's when she discovered that one of the boards on the table had actually been broken and bashed in. Somebody down there just was not happy that we had walked into his woods. And we only walked in about forty feet, we didn't go far, but I think that evidently was too far.

Back to Ammi

Yesterday, when Rita and I went out, she stayed closer to the edge of the woods, and I really wasn't going to venture that far back in there but my curiosity can get a little better of me and there was some stuff I wanted to look at. I got down in the area and took me some pictures and I didn't feel too welcome. Just a little uneasiness, so I decided I'd come out, call the dogs. When I started walking back out, that's when it hit me in all the large muscles in my legs and in my rear end. My legs pretty much just buckled. I just kept going. I got real disoriented. At one point I left the little path and was walking through brambles and Rita had to holler at me and let me know which direction to back off to get back on the path. It was very visible in there. I could see her, she could see me, we could converse pretty easily. I wasn't that far from her. But my legs didn't really want to work. The best way to describe it is, I've lived in the northern states and if you've ever tried to walk through a three-foot snow drift, or running in water, that's pretty much how it felt to try to move them legs. By the time I got up to where Rita was, it started easing off a little bit, and the thing that was really weird was when we hit the edge of the woods, as soon as I stepped out of the woods into the field, it disappeared, the whole thing just disappeared.

There are so many little things they do that have nothing to do with collecting hard evidence of them, but is just fun and lets me see and hear the behaviors. That is my main focus here . . . I want to know all about them. Having the young ones so close here is a delight most of the time. In the same way for me as I loved watching and hearing my own kids play, and grow up. I tend to get comfortable with them being playful, and am reminded with a not so gentle zap, or growl, that I am too close, or somewhere they don't want me. Then I back up for a day, and get right back out there and try again. I am trying to come up with different ways to do things, to see what works.

When my grandchild was here last year, at age six, she used to walk near the woods and talk to herself constantly. She would take her toys and play there too. I would be gardening, and ask her who she was talking to. She would say the hairy people in the woods, or the hairy kids in the woods. I thought she had a good imagination. I didn't know anything about the Bigfoots back then.

She was walking near the "nest." This is a long strip of land, next to the swamp, and bordered by the wood area, until it hits my yard. It is made up of loads of evergreen brush, vines, and stick add-ons. There are two and in some places three "screens," along the side towards me. These are made of vines with added sticks, logs, brush and leaves. They have been built up even more near my house, daily. The one near the back of the pumphouse used to be easy to take pictures through, and is now almost a solid wall of debris.

I would say the nest ranges from ten to thirty feet wide, in places. It is at least fifty yards long because that structure is not all vines there, it is also laced with a lot of small sticks and branches.

Other people, habituators that I talk to who do this at their places, and do have kids, say that the Bigfoot are always watching the kids play, and they get a lot of the best pictures when they are doing that. They are the ones who told me the Bigfoot love the stuffed toys they put out, and will keep them. Other toys are moved and scattered at night, or taken, but are always returned eventually.

So I am going to try that. I will be leaving them near the feeding and nesting area, and see what happens.

I was also told they like plastic flower pots, and that explains why mine are always ending up in the woods here. I do have three boxes of old toys in the barn. And they get scattered a lot, and stacked on each other. I hadn't thought of it being them. I just blamed the dogs.

January 23, 2008

Tonight, they have been playing games on my roof, since it got dark. The porch roof is right by a tree, and then they climb onto the house roof. They've tossed rocks and a branch down my chimney, in the last few minutes. I went out into the cold twice, but of course they are gone, or ducking over the roofline.

My dogs have been going nuts, and judging from the new branches in my walkway they must be throwing at them too.

Last time they did this, they tossed pieces of bricks down the bedroom chimney and woke hubby up, then a handful of live lizards came down . . . freaked him out . . . he hates lizards . . . made me catch them all.

This has made me think of something, though. They do seem to be very active in the colder weather. Thinking back on my days before I knew it wasn't humans trying to mess with me here, it happened more in the colder days than in the summer.

So far, it has been smaller ones up there. I think the juviniles play those games. I hope none of the big guys ever take a notion to go up . . . could damage my dome.

I have floodlights on the house, and turn them on when they climb up there, or pound on the house. I turn them off after a couple hours, and it has stopped them before, did tonight also.

I do tolerate the tapping on windows and at the electric box. That is done more gently, and for attention, so I treat it as such, and just tap back from in here, or talk to them loud enough for them to hear me. It is a game that is played for a while, then I usually have to be the one to stop it.

My husband refuses to believe, as in, everything can be explained by deer or owls. Yep, we had a fun debate about who dropped the bricks and lizards down the chimney. He has heard them, and refuses to go behind the barn at all. He used to say there was nothing out there, but now he slips up, and he talks about them some. He just hates the whole thing, and would rather I do what his mom's family did when they were living here in the past, and just ignore and deny them. He really only gets mad when his mom starts in on him about me studying them. But she doesn't say they are not here, just that I shouldn't "mess around with things I don't know about."

When Rita came and brought the camera here, as soon as she left, he started to move the couch from in front of the window that faces the swamp. I asked him what he was doing, and he said, "I thought you could get the best shots from here."

I said, "Shots of what?"

He got mad and said, "I was just trying to help!!"

I just laughed, and helped him move the couch.

February 4, 2008

I had an interesting day here. I had a half-frozen lizard on my kitchen steps, rare, because they usually are under something in this weather. It was thirty-five degrees out today. And this little guy was so cold he couldn't move, thought he was dead at first.

Hubby had just come in that same door, not long before I went out and found the lizard. Maybe it was a gift for him?

I saw in front of the door the bag I'd left chips and Cheetos out in, near the woods, so I went to check the pumphouse, where it had been hanging, about six feet up inside. Gone from there. I went around the pumphouse to look, and found some small footprints, about the size of our six—or seven-year-old grandkids. One was almost perfect, so I got pictures. Still don't have casting materials here.

Went and picked up the bag, and took it, and popped the lizard inside, so I wouldn't be shunning a "gift." (I later snuck him back out . . . and tucked him under a pile of leaves.)

I put the bag on the big table, behind the barn. While I had my back to the woods, the dogs went nuts towards the woods, so I turned to see one of them chasing after something that I couldn't see, into the nest area. The pups were looking at something hard, just under where the stuffed teddy bear is hanging, so I went to check that out. The ground had been dug up and dirt was all over on top of the leaf litter, so I went in closer. Ewwwww! A fresh deer leg, no skin, half of a pelvis, the thigh, knee, and part of the leg. Still all connected, and half full of meat, the other half chewed off. The hoof had been broken off. The whole thing was still warm, even though it was freezing out.

This was only about five or six feet from where I had picked up the bag, so I couldn't believe I hadn't seen it then. (Later checked the pic I took of the bag, and nope, it wasn't there then.) I don't want to turn down "gifts" . . . and know they have "gifted" other habituators with deer, so I made a big deal over it, and giving it to my dogs, who wouldn't touch it at first, but then one finally grabbed it and ran off with it.

I thanked the wookies out loud, and headed inside.

I want to study behavior, share what I am doing so others can have some idea what it is like to do this, and mainly have me and the critters here, left to live in peace. I do bring some people in, for the study, who have the skills needed to try to get some evidence, but it is done quietly, and not in crowds. The BF react to different ones, different ways. I note that, and try to figure out that also.

They are most active if I am alone and tend to hang back more if even my hubby is out there. They do favor my son being out with them, at times, but he has a pretty strange way of playing with them, and will go out alone to do that at night.

I get more good reaction with the younger ones, and the mammas and day ones. The night ones scare me at times . . .

My son gets reactions at night, good and bad, but he is twenty-two and "bullet-proof," so he takes more risks than I do, and enjoys doing so, or can at least run back inside quicker.

I can, however, sit quietly for hours, and just listen, and look for them. I can go out and sit close to the area, and sing, hum, whistle, play "Peek-a boo," use myself as entertainment, like you would for a toddler, and make a total fool of myself, just to try to get a reaction from them. And then sit there even longer, ignoring them, to try to make them get me to react, which gets me movement, and vocals, and maybe a picture. That is what I do the most, and it seems to be working, as far as getting them used to me, and accepting me being there some. I push it too much at times, and then have to back off a few days. It goes on like that.

I have learned that if I ignore them totally, don't raise my blinds, and don't go outside at all, by around noon or one pm, they will come up and tap on a window near me . . . easy . . . . or jiggle the tag on the electric box, next to where I sit here, on the computer.

I always have rewarded that, by pulling up the blinds, and taking them out a treat. They know that now, so it is a regular thing, if I do that. No, I am not going to jump up and run snapping a camera, it would totally mess up what little trust they have, which is more important than a "great picture right now." If I can build the trust more, I will have much greater chances for a very good close-up shot. I have that kind of patience.

March 17, 2008

Today, I spoke to my neighbor on the other side of my swamp.

He had called me over there to take some pics of his pups . . . he has no computer or camera. He wants to have me help find them homes. I went over and took the pics, and was getting ready to go, and he asked me if I had ever heard or seen anything "strange" in the swamp, since I had been here. I told him, "Yes, plenty," and he went on to describe all the sounds. He has had two sightings, believes they are "a people, and not devils," and described being paced, and zapped. He asked me if I ever heard one whistle a "funny little tune," and when I whistled the *Kill Bill* song, he said, "THAT'S IT!!!"

I told him I had taught it to them, and he now wants to see if we can teach them "Amazing Grace," if we work from both sides on it. So I'm game for that one!

Family B: Oklahoma

The speaker is a Native American woman in her late forties. She is a medically retired police detective, with a bad back. Her daughter and grandchildren are frequent visitors.

1954

One of my very first memories of a Shadow Person was on a foggy morning. The time of year was early summer. I was about four years old. My brothers and I spent hours in the woods near Ma's house. In those days families didn't worry about kids being taken. Hours were spent outdoors.

I recall Ma was cooking breakfast, as I slid out the front door headed for the woods. The woods were thick with locust trees, grasses, and moss. This particular morning the fog was thick. I wanted to see what it looked like in the woods. As I walked up the road to the entrance, I saw them, all four of them, four tall shadowy figures. They looked like an Indian family. The difference was they were all gray. The father and mother were standing behind two children. Their faces looked Indian to me, like Ma's. They were the tallest people I had ever seen. I stopped a few feet in front of them and looked at them. Their expressions didn't change as they looked back at me. To this day I still remember it felt like a dream staring at them. They had grayish hair from head to toe. There was no hair on their faces, hands and the bottoms of their feet. The hair on their bodies was probably two inches pretty much uniform in length. It lay flat against their bodies. The hair on their heads was more ruffled. As if it needed to be combed. They stood for probably only seconds. The largest one, the father I assume, turned to his right and walked away with the others following. They never made a noise. Everything was muted. They disappeared into the fog and the woods.

I walked up to where they had been standing. I remember wondering why the edge of the bank there was crumbling. I was never afraid while standing there looking at them. I was more curious than anything else. I felt lost.

Then, I turned and ran all the way back to Ma. When I entered the house I was confused. I remember telling Ma about them. She told me those were the Shadow People, leave them be. She was very adamant never to follow the Shadow People. There was a possibility if you did you would never return. Ma said that children that did follow them often get lost and confused.

For several years we ran and played in the woods always under the watchful eye of Ma. She never allowed us out to play until she checked for signs. She always checked the scents on the wind and the tracks in the sandy road. Once in a while she would not allow us out to play until late morning. She never told us exactly why, just wouldn't.

Over the next few years Ma would occasionally mention the Shadow People. After I, her persistent grandchild, insisted on it, she explained they were ancient people. They didn't bother anyone nor did they want to be bothered. There were several tribes of them. They were all people, just different than us. Ma explained they avoided people like us.

One summer day, Ma called me to her and read me an article from Readers Digest. This article was about some men that had filmed a Bigfoot. [The Patterson Film.] Ma showed me the tiny pictures in the article and told me that was a Shadow Person. "Never bother them, it's bad to hurt one," she told me. "Leave them be." They wouldn't hurt me, I should just leave them alone.

When asked to share with others the knowledge I have obtained over the years, I realize it's not as easy as it once appeared. It's complicated to talk about the Shadow People. Through the years they have been labeled by so many with every imaginable tag—mythical, imaginary, monsters, monkeys, apes, cryptids, hominids and of course unknown primates. There have been many who have chased them, begged and pleaded with them to appear.

I do not care if you question my credibility. I do not have to prove who I am or what I am. I have simply agreed to share old and new things. I cannot prove to anyone what I am sharing. You will either believe or not believe. Whatever you decide will be up to you. All photos and video were taken by myself or my family.

I only care that the Shadow People remain in peace to rear their children and live their lives. At this moment it appears many wish to murder one or a family for science. Killing one is murder. These are the ancient people of the world. Call them what you will, they have survived much and adapted to remain in the quiet areas of our world.

As I grew, every few years my path crossed with the Shadow People. Each time it was a different place, several years apart. Each one appeared different in color, shape and size. Each one was in an area I would characterize as the fringes of civilization. Never have I been threatened or harmed. Each time the Shadow Person was the one who disappeared into the shadows.

Years back, after relocating my family to a rural area, I discovered that those Shadow People were again close at hand. They are as interested in us as we are in them. They will if the opportunity arises peek in your window. They will stand back and watch you. They have never approached any member of my family or myself. I do talk to them. I always respect the personal space that we each need to maintain our relationship.

On many occasions one of the kids will be outside and come running in asking, Did you call me? I have been working in the gardens or sitting outside and have heard a child's voice call "Mom." It sounds just like one of my kids. This has happened many times over the years when I was home alone. They are known to knock on houses both day and night. Why they do this is anyone's guess. I suspect it's a form of communication, checking for a reaction.

I will never be able to answer for myself or for others all the questions about these wonderful, gentle creatures.

Summer, 2007

My nine-month-old grandson Squirt was sitting in the sandbox at the time of "the monkey-chase," as the kids called it. He was reaching and laughing at the bushes where they burst out. It was his behavior that got my attention at first. He was cooing and laughing as he reached out in that direction. I thought wow he must see one of the cats.

A split second later the first one burst out of the brush and ran bouncing across the yard. I am convinced Squirt had spent several minutes watching the monkey-kids in the edge of the brush. We call them that because they look like the drawings from the rise of man or the hobbits.

Actually I knew they were coming around for years. I just kept it to myself until about five years ago and told my husband and dad. I finally had good prints to show them.

It was when I realized that so many want to kill them that I stepped forward and met others who were looking for them. I believe them to be hominid and not apes. Sorry, that's the way I was raised. I have gradually begun sharing things I have seen and heard with others. It's hard to learn to trust anyone. I decided it was time to teach the kids and grandkids the old ways also. As I was taught.

This summer just brought it all home. My little five-year-old grandson is the one, besides the nine-month-old, who can spot them hiding in the brush when no one else can. Taking pictures of the areas he says they are in generally shows shadows shaped like humans. I don't want the kids to fear these guys, I want them to learn to respect them.

But back to the monkey-chase. Just before dark, I saw three of them. The grandkids say there were five altogether, plus the babysitter. She was about the color of this little guy in the picture I took, maybe a bit more auburn.

I saw one very dark one, mouse-colored, and one red. The grandsons said there was actually three very dark ones and the two lighter ones. They were everywhere in a matter of seconds It was one of the most amazing things I have ever had happen. They burst out of the bushes, ran all around us, and the kids gave chase like a game of hide-n-seek. They won hands down. Then they were gone. They were between two-and-a-half and three-and-a-half feet tall, rail thin and fast as the wind. Along with a female about five-and-a-half feet tall hiding in the shadows and holding a baby. The babysitter kept reaching out from the brush as they would run by. She ran up and down in the brush. The most amazing thing is they never made a sound. None of them. As soon as it was over my grandchildren put out a pizza and P&J sandwiches. Which disappeared right after dark.

I have lived here sixteen years and they have always traveled through back and forth a couple times a week. Year-round. This year the lake came up and they hung around a lot more than in years past. However, this year the grandkids converged as a group more than usual. Generally the kids are here one or two at a time. Plus the baby was here a lot.

I have always fed the wild animals. It's a hobby as much as compassion for the creatures the Creator has given us. This is a way of giving back to the Creator. We have so much and the little fur balls have so little. The more you feed the more they will come.

Years back we built a small garden pond, more of a frog sanctuary than anything. Stocked it with minnows and goldfish, which were promptly eaten. This was an ongoing pattern. Stock the pond and see how quickly the fish disappeared. A garden pond can be a great source of entertainment, also much work, depending on how you wish to handle it. Building an ecosystem takes time.

As each spring came and went more was added to the pond. It was during this time that the prints began to appear around it. The first bare prints measured fourteen inches by five-and-a-half inches. They appeared on the garden path beside the pond during the winter. They also appeared in the pond. As the footprints appeared the fish disappeared.

Each year the fruit would ripen. Just as it was ready to pick it would disappear overnight, never in the daytime. The bare footprints were different sizes. They ranged from a mere four inches long to nineteen inches long. The amazing thing about footprints: often they are beside the paths, in the grass, as if to avoid detection.

Since the beginning of the year I have started to organize and keep a brief record of the Shadow People in my world. After the ice storm last winter we emptied two freezers of food. No idea how many pounds were placed by the habitat area. All disappeared overnight. During this time the bird feeders would often empty overnight, also, as would most of the feed on the ground.

About a month later, during the night, someone jerked on the backdoor. Several times, this actually broke the bottom hinge off and sprung the door. My thoughts were that the Shadow People were hungry and knew that there could be food in the freezer. Over the years we have actually lost food from the icebox and freezer, never giving much thought to it.

As the ground began to thaw there were faint tracks appearing. They once again were different shapes and sizes. We purchased a trail camera to use and play with. It isn't for serious hunting, it's merely for the enjoyment of catching our nighttime visitors. I was hooked on trying to see more.

As the winter began turning to spring, I was recording audios and some videos, never catching a Shadow Person on film. Although

I believe they did walk by chattering as in a conversation. This audio was shared with several others, including researchers, for evaluation. I have hope that someday there will be an answer to what sounded like a foreign language being spoken in my yard at two am. I have had cameras moved, picked up, set down, and turned over. I have had only shadows to show for hours of video.

We have some neighbors. I don't know most of them. They moved out here from the city and tried to make country like the city. I don't think country agrees with them that well. They are putting up cameras lights and fences. For whatever reason!

I suspect the neighbor right next door has seen them. He is an older guy and carries a big pistol on his hip and a rifle slung on his back to walk around his yard. There for a while he creeped around in a ghillie suit. Built a brush blind and just crawled around the yard. He had an alarm that was hooked up to an air raid type siren. Imagine what fun that dang thing was. Every few minutes it was like WWII around here. Some of the other neighbors finally called the law on him.

This year, I began putting out snack food for the Shadow People. I tell them when I am putting it out that it is for them. They often are hiding in the woods nearby and I am certain they can hear me as I talk to them. This food is not in the same areas as the critters' food. I separate the food out of respect. I would not serve a guest food from the dog dish nor will I share snacks for the children of the Shadow People with the raccoons on the ground. My quirk. I was taught to respect them. At first they didn't touch the snacks. In fact nothing touched the foods. Which was very odd. This continued for several weeks. Then one night the snacks were taken. An audio recording of this indicated during the rain the container the snack was in was lifted out of the tree and returned. This container was approximately eight feet from the ground.

Thus began the game of "snatch the snacks." Certain nights the snacks would be taken. Soon, I began attempting ways of catching the Folks on video. Never have I gotten more than a shadow moving or sounds that I have come to associate with the Shadow Family. On occasion I will have one container moved to another tree. Sometimes it will be higher or lower in the tree. On rare occasions, I will need a ladder to take the container down to refill it.

I also leave gifts of beads, toys, balls, and other interesting objects. Sometimes, toys and balls will disappear for a period only to reappear months later. Sometimes the reappearance will last for only a day or two and then the objects will disappear again. The working assumption is that it's the Shadow Folks moving the items. At this point I have doubts that possums and coons want trinkets.

The toy house is a playhouse filled with abandoned toys not played with any longer. As the main house overflows with toys they are crated up and moved to the toy house. Only to be forgotten. This year I checked the toy house and found that most of the toys were no longer there. On occasion, I find small toys scattered over our acreage. Based on the behaviors I am observing, the Shadow People never take anything they know belongs to some member of the family. It appears they only take what is ignored. This not only includes toys. At times garden tools will disappear and reappear months later in an odd place.

While playing with my grandchildren one evening, I noticed a change in the birds singing. There was a snapping of tree limbs and soft noises in the brush. I started snapping pictures within the area. There is no panic feeling. No fear. We continue about our business. I firmly believe that no harm will come to anyone who has a relationship with them. I will say never leave small children alone. Not because they will be harmed. Children can and will try to follow these guys. A small child in the bush is never a good thing.

It is truly a blessing to have even a glimpse of these wondrous people. Now, that being said, there are many reports of these creatures being frightening and dangerous. Yes that is probable and possible. I am always puzzled by these reports. What makes a creature, any creature, become aggressive? What makes a killer a killer? I will not dispute anyone's claim of aggression by a Bigfoot. I refuse to argue this point with anyone. I suspect that if there is aggression there is a reason for these actions. My thoughts are perhaps that aggression is due to getting too close to the young. Most animals, including humans, will protect their young by whatever means are at hand.

I have never witnessed aggressive behavior from the family unit that often visits us in the night. My family and I spend time out there at all hours of the day and night. Never while we are outside do we feel threatened. Watched, yes. I don't know how to describe it other than just that. Somewhere close by someone is watching.

It does, however, seem that the children and women more often have that feeling. There are possibly several reasons for this. One, the creatures watching are more interested in or entertained by children and women. Or possibly they withdraw further back when it's males about. I personally think that it's because men have a different air about them and are often the ones in the woods with guns.

Often the feeling of being watched is accompanied by soft sounds of movement. Sometimes sticks breaking or being pushed aside. It's at these times I try taking pictures of the area the noise is coming from. I have tried to school myself not to put a camera up to my eye or make a quick movement. Looking at photos taken at such times, I feel it is safe to say that most often what you will see in a photo is partial facial features. The eyes are what I notice most often. After finding an eye then I look for the other features.

I am not making a claim that the Shadow People live, reside or dwell on or near my property. What I do know is that they wander through at times. There have been several occasions in which we have gotten glimpses of them. Never the prized face-to-face.

There are times, especially at night, when I can tell the Shadow Folks are about. What can be said is if there is a whistle or click and it's verbally answered it will become quiet. Almost like an embarrassed quiet. The Shadow People have had millennia to perfect their camouflage. Like any creature, they adapt to the area they live in.

Until the last few years I had never tried to have a relationship with a giant creature. If you were to ask us to describe how we feel, knowing big hairy guys/gals are wandering the area, well you would get as many answers as there are people in the family.

The problem with a relationship with the gentle creatures of the forest, especially the larger ones, is a basic fear factor. It's built into each and every one of us. There is the fight or flight reaction. I would tell anyone who wants to know more: take it slow and easy, there is no other way. This is a trust that slowly develops over time. We are talking about a creature that our Creator placed here with us. This creature to my way of thinking is on the same level as those of us without the constant hair shirt. My ancestors believed that to a certain degree if they come forward to you they have a reason, such as an immediate danger to you or your family.

At this point in my relationship I believe that may well be the case. They remain elusive and silent. Habituation any way you want to present it is in fact a relationship. Started by one side or the other, for whatever reason. If you want a relationship with the Shadow People, remember they are intelligent. They are masters of hunting, camouflage, and probably fishing.

I see no harm in gifting, but boundaries need to be established in habituation for the peace of mind and comfort of both sides. Never give what you do not want to be taken. Never set boundaries that you cannot live with. Remember, setting them may not work with intelligent creatures such as the Shadow People. They have had thousands of years to roam about. Just because you don't want one at your back door doesn't mean you have not encouraged this behavior. Think before reacting. Overreaction will not help with any relationship.

I didn't start out to form a relationship with any animal in particular. I certainly never expected to have a tribe visit me in the middle of the night. I do feel that this is a very special gift from them to me that they have chosen our home as a respite at times.

Here's an example of a reaction that I can share. Recently in the middle of the night I was awakened several times to the sounds of scratching on the windows and sides of my home. Annoying to say the least. This continued for several hours throughout the night. And what did the guys want? I have no idea. The scratching noise came to a halt with what sounded like a firm smack and a wounded baboon yelling and running away from the house. What I believe it to be was an errant child "pranking" and Mom becoming tired of this behavior and correcting it.

My reaction was getting up and actually taking the time to try to guess where it was coming from. Each time I turned on a light the noises stopped. As soon as the lights were turned off the noises began again. Did I see Bigfoot standing looking in the window? Nope! I do believe it was the family wandering about. I had left gifts of fruits and nuts for them. I believe they were about because of faint odors that had been detected by my family earlier in the evening.

Our youngest is now eighteen months old. He has a thing he does every chance he gets. He goes to this spot and talks to the window and/or the bushes outside. He doesn't get into anything or bother anything here. He just sits and talks in baby talk to whatever is out there. He waves to it and motions for whatever to come to him.

At first I thought that it was a fluke. Then I started paying attention to him. He does this quite often. He acts the same way he did the night the monkey kids blazed through the yard. He sits and talks and motions to something. He did that day at nine months old also. Even stranger is when I approach he stops and acts like he is caught doing something he shouldn't. He will not, as long as someone is watching, continue with the baby talk and motions.

I did notice one of Squirt's toy guns is laying out there on the ground. There is a well-worn foot path outside of that window. The interesting part is we rarely us it. The path goes right along the side of the house and out back to the brush and chicken house now. I Googled here the other day and you can actually see the foot paths. One of them to the chicken house. Which we have not used in several years. Over three years to be exact.

As I started actually trying to communicate in some manner with the local Shadow People here I tried many things. I have read many on-line conversations with many ideas on communicating with or "baiting" the big guys. A lot of ideas I dismiss as silly. The silly ideas generally are people who assume that these guys are of lower intelligence than us and must be treated on the level of a child, or a pet.

I fall back on treating the local Shadow People as I would want to be treated. I hate pushy know-it-alls. I avoid them myself. Using how I feel about neighbors and being a neighbor, I started with food and little gifts, which were often taken. I began to add other things. Pebbles, feathers, ribbons, all of which have been moved around, relocated or taken.

Then, one day, I wondered how roses would be received. I picked several rosebuds of different sizes and colors and included them with the other items. I found the buds the next day nearby where I had placed them. They were taken apart petal by petal, and left.

A few days later, I awoke and went out to have coffee. I found a trail of pink rose petals to the area where my chair was. At the time I thought, hmmmm, that is odd. Later, I questioned my family about this. No one had done it.

On another morning we found several red roses lying on top of the car.

Several weeks after this, I again found a trail of rose petals from the door to the patio. All in all on three different occasions I found a trail of pink rose petals scattered from the front door to the patio.

Our household is taught: "What is done to the least of these is done unto me."

Family C: New York State

Jerry is a thirty-year-old man who lives with his parents. He has been having experiences with Sasquatch since childhood. I have been corresponding with him since November, 2006, and am able to visit him on-site twice, in June and December of 2007.

May 1985

He was six years old when it happened.

"When I met them for the first time I had peanut butter and jelly sandwiches," Jerry told me, "and as I remember it they stayed close to me but divvied the pieces up between them. Remember we were way back in they woods so there was nothing for them to run and hide from it was just me. When they finished what was my lunch but gave to them instead they wanted more but that was it. So [the main male] Miget and his woman—I call her Momma—seen that I gave them my food and fed me some of theirs like some berries we went foraging for. They tasted good from what I remember but I knew I was going to have dinner waiting for me when I went home."

For the next twenty-four years, Jerry has kept up consistent and loyal contact with this same group of Sasquatch. Because they live in a severely limited tract of land, boxed in by suburban residences, he has steadfastly kept their secret, knowing that any leak could cause their ouster, if not their destruction. They have subsisted here, in fact, since before Jerry's great-grandfather first interacted with them in the 1950s, and these accounts were passed down through the generations.

The more he has come to trust me, through our correspondence, the more he has felt comfortable in sharing with me of his past with these *people.* "I like to refer to them as people, not animals." He thinks there were "many races of *homo sapiens* roaming the earth at one time but many of them went extinct. I just like to think that this is one race that slipped through the cracks and remained on the wild side of things."

He has also long known that he'd like to find a trustworthy researcher to take along to the site, to help him to officially document their reality (without exposing their home territory), to establish it definitively within mainstream science, and thus to get the wheels turning toward legal protection for the species.

I agreed not to breathe a word of their location, and I have not. That they have managed to survive without discovery in a constricted stronghold of forest—a forest that, albeit thick, comprises just seven square miles—speaks better than any other case I know to their uncanny stealth and adaptability.

"I mean lots of people know they're back there, even my great-grandpa knew about them and told me about his experiences with them before he passed. He even called them his friends. I just think it's kind of weird how I got to know the same family I like to call the Migets. And it's not certain that they just stay in [name of forest]. I'm sure they go other places, but I know they like to stay close to the stream for water.

"I can't say how many are back there but there are at least five or six that I know of. There can very well be more because they are too unsure or bashful of us men. When I first met Miget, the patriarch or leader, his father was still alive and he was very old but didn't look like it through his actions. They are a very peaceful group of people but very curious of us and at the same time even more scared of us. They are caring of each other and protective of things they take care of. They are even more curious of our foods and simple luxuries like hair combs.

"One time when I was fly fishing, Miget's boy—I call him Junior—walked up behind me and scared me because I didn't hear him till he was right behind me and he was all fuzzy because it was March so he still had his winter coat, but he was only four to five feet tall I guess you could consider him a little guy. He was very uncertain and a little offensive but very curious and just like that he disappeared when I turned my back to check my line.

"The few times when they vocalized for me, I could make out some of our words strangely enough. I hate to sound like a delusional crazy person with all the info I've come forward with but it's what I know. These are wild people we're dealing with and with all their humanness they're still very wild. It's kind of spooky when you think of it or when the situation occurs.

"If someone don't have the right look they won't trust or understand what you say through the simple sign language along with slow talking. They also understand when you speak like them it comes natural when the moment arises, it's weird. There's no need to practice. The feelings they feel conjure up in you like a link to our common past.

"One time, I brought my friend back there when cutting school and we were walking about a quarter mile back and I looked to my side and there was Junior. Just sitting there with a grin on his face looking at me like an old friend being reacquainted. My school friend started shaking uncontrollably and ran back to school. After Miget came into sight with his brother we had a visit for a while and I went back to school myself. When I ran into my friend in the hall he was so shaken up he forgot what happened. Which is good because I think that happens to a bunch of people because it is so unbelievable that their brain doesn't register it. Just like the Indian folklore they say that they have magic and control you.

"I'm not really out for self-glorification, but want the truth to be known before it's too late or introduced in the wrong way. But you want to know what my theory is? These people are going to be here after man is extinct. They'll be using our structures in a simplified way. Like a highway overpass would be their idea of a home, for instance, and so forth.

"Among the Migets, the female has something beautiful about her. You can tell she's a woman by her figure of course plus the way she walks and by the look in her eyes, it's gentler. It's definitely different than the guys. That's once you get them to feel comfortable. The guys have a build that will make the world's strongest man look like a little girl. The ladies are built and strong too but different if that makes sense.

"They are like us but they groom each other and they got wild eyes if you look them straight on at first, then they relax. They're like a married couple, they bicker and everything. Plus I didn't want to share this because I didn't feel you would believe me, but they had a baby. When I was a kid. I seen him the second time I was taken back in the woods. I might be the only human to see a baby Bigfoot. Very cute, he looked like a newborn maybe a foot and a half from head to toe or two feet maybe.

"They had him in a piece of cloth they scavenged from somewhere. I forgot to tell you when I seen Missy [a younger female] the first time, Miget came out of the woods about fifty feet away and called for her like *come on let's go you*, that was the tone. So there's Miget, Momma, Sissy, Junior, and Little Brother. Plus Miget's brother who I call Uncle Bugout, because he loses his temper. I'm just amazed how those little ones do alright in the winter. This particular family is lucky they have our pollution and waste to pick from for a baby blanket if that's what you want to call it. It makes you feel bad when you think how poor the are but happy.

"Oh, Miget. He was a very big boy since I first knew him. I can remember standing in front of him and having to tilt my head totally back towards the sky to see the top of him. It's still like that today. Miget has been at least ten feet for as long as I've known him, and three or four feet wide, a true powerhouse. He can clear a path running full speed."

June 23, 2007

Jerry and I meet on warm June afternoon teeming with goldfinches and deafening insect life. Our plan is to spend the day and the night inside the forest.

I must confess here that even after months of copious email exchanges, of apparently sincere testimony from this man, I'm still feeling only about 50/50 on the question of his ultimate credibility; after all, what he is asking me to buy into—in terms of his childhood experiences—seems far too good to be true.

Add to this, now, the fact that although we are meeting at his current home here in Upstate New York, Jerry spent his youth (and knew "the Migets") *downstate*. He claims to have made contact, over this past winter and spring, with yet *another* family group, hundreds of miles from the first.

This fact does not, however, strike him as remotely strange; his contention is that these people live in most forests, but that human beings do not, as a rule, have the foggiest notion, much less any aptitude for outreach.

Jerry is just about my height, six feet, heavy-set, with bad knees that cause him to sway back and forth a little when he walks. He works in a machine shop, where he puts together transmission housings. On weekends and occasional days off, he is free to spend quiet hours inside the forest, just two miles from town. What strikes one first is how soft-spoken and down-to-earth he is; nothing mystical or rash passes his lips, and his body-language is understated, facial expressions even humdrum, the opposite of what one might expect after hearing his mind-boggling stories.

As he leads me through tall grasses on top of a high clearing, he's more interested in agriculture than in Sasquatch, pointing out the various farmers' fields, telling me their names, and whether or not they've fallen behind schedule in the season's first haying. He clearly enjoys the look of these swatches of land arrayed below us in the valley.

On the downslope, we approach and pass through the tree line, leaving the goldfinches and bright sunlight behind. It takes us no more than ten minutes, once inside the woods, to make our way over roots, fallen trees, and along informal fernways to Jerry's hide-out, a simple but effective lean-to he's constructed out of sturdy branches and pine boughs.

The very moment we arrive—and I mean the *very* moment, even before we can sling off our backpacks—a single, distinct wood knock pierces through the thick trees, coming from our west, maybe seventy yards away. Jerry looks at me: "I told you so."

I have to laugh, from pleasure in the hearing, and because he certainly *has*, in detail. We set up a sparse camp—thermoses of coffee, collapsible chairs, candy bars, a candle set on a rock—and then spend four hours alternately talking and meditatively listening. It's breezy, my heart's bruised from a fresh, treacherous breakup with the mother of my child, and sunspots swirl over the forest understory.

What follows is a blend of Jerry's words from our conversation this afternoon and from his emails to me over the past six months.

I start by conceding that during all my BFRO Expeditions, and certainly in my solo time in the woods, I've never gotten such an unmistakable overture as the wood knock we just heard.

"Maybe you're not listening enough once they know you're there. They'll give you one really soft knock. That means you're in their sight and you don't even know it. It sounds crazy but you have to become an animal in your head.

"Remember, you're going into *their world*. They have super senses as it is, so you shouldn't have to over-extend yourself to get their attention. Keep it simple. I feel the BFRO might be too loud when they're on expeditions. When you think about it they already know we're reckless and loud as a race. That's why they stay away altogether. The researchers stay in a crowd out there. Why fill the woods with a bunch? Simple sounds, little whoops go a long way too. That's why they aren't yelling out more as well. They don't have to, they already got great hearing. I'm the kind of guy that sets off in the woods by myself. I don't like talking out here unless I try to talk with one of them, like last week when he threw a rock at a tree by me. Then I talked calmly but excited. Yup, keep it simple. They get spooked about people to begin with.

"They can feel when a person's spirit is good and means good, that's why I've gotten this far. They're just watching out for themselves right now, that's what's happening. You know this all seems simple but the biggest aspect of this besides them is to share good information for their sake and not to let people see them the wrong way but to love and respect them as the fathers of the woods."

Again and again, he stresses the indispensable importance of sincerity.

"Once they know who you are they aren't offended or threatened. I think I just lucked out or maybe I have a peaceful look in my eyes or maybe it's because I am a big guy or I look and act like them, ha. I do enjoy spending lots of time in the woods too, maybe I'm being checked out from a distance and their curiosity brings them in.

"I have to tell you about my experience in the Poconos. When I lived there ten years ago sometimes there would be a thump or tap on the wall of my house on the wall behind my T.V. at night. I would be wondering what it was so I went outside one night with my dog and out of the woods came a seven-to-eight-foot teenage male. My dog Jack got between us, and it almost attacked him but I walked up to my dog and started petting him and saying good boy, with a calm voice. The male looked in amazement as Jack calmed down, then I started to play with my dog and he seemed to understand. I could swear that he said "Jack" too in a pleased voice after I introduced him."

This is the type of thing, of course, that stretches one's credulity. Jerry relates the anecdote as if he's telling me about an innocuous street-corner encounter.

Winter 2008

Last winter, he wrote me near-daily updates on his careful process of outreach to the Upstate locals, when he first established their presence.

January 12, 2007: "I had a gnarly day out in the woods my friend. I went for a walk four or so hours ago and just got home. I brought my homemade club for wood knocking. I was knocking the whole way back on that road I was telling you about where I seen the bear, they're asleep now I'm sure. But I also know there are Migets in the area because of the deep woods and past experience.

"Well, I went down this path leading deep in the woods. I gave a power pole a good crack before I left into the woods. I wasn't too far when I heard it, a huge thud coming from across the valley in a deep patch of woods I could pick out. Along with another crack far deeper and more powerful than I could ever make. Well I kept on my way out there into the wilderness. I kept on finding things to beat on and finally found a nice rock with a good flat side. I stayed in one spot so I could listen with a vantage point over two valleys to hear and see over the woods and farm fields, and so they could hear me. I was getting responses now only with far-off knocks in the woods. I was knocking my club on the rock and gave it a rest. Then I couldn't place it but there was definitely something on two legs walking around me somewhere with big feet. I could tell from the sound the snow made when it packed, it was so cool. I made contact, not how I wanted though.

"I wanted to stay out there and make a visual and even a close-up encounter. Oh yeah, I also seen some foot prints on the seasonal road on my way back. I couldn't say how big or how fresh because they were wind eroded, but they sure did look big."

I pointed out, by email, that most people wouldn't be able to stay out for hours with their face exposed.

"No I like the cold. My jaw was hurting when I got back. With the wind it was like a sandblaster out there. But about the knocks.

I couldn't believe it at first. I figured someone was chopping wood. But that was a thud. I also heard a vocalization, a whoop with a little monkey chatter. It didn't last for more than a few seconds, twice.

"At first, the knocking came from my east then after a while it came from my west. Which means it was trying to locate me. Then the footsteps packing the snow put it in stone for me. The steps sounded like they were within fifty feet of me. That means he or she could see me but I couldn't see them. In that state of mind looking around to figure out where the sound was coming from, you could easily overlook a perfectly camouflaged figure in a dark woods, plus I was down in the ravine. It might have kept back once the wind really started up and blew my scent toward it and it figured out I was a man. Otherwise it might have just walked up to me until I came into its vision. I was hidden in the ravine good enough to not be seen. There were snow banks tucking me in."

January 21, 2007: "I spent about, oh, three or four hours out there today and just got back five minutes ago. I headed down the hill on the dirt road and went east by the pond. I was only on that path a few seconds and seen a set of tracks come out of the woods and head down the path, I couldn't tell what direction, they weren't fresh. They didn't have a big stride, so I figured it was a young one. If it was a human what were they doing out that far and why coming out of the deep woods like that? Then again what was I doing out that far? So I said to myself, this is the spot to try today. So I took a good swing with my Bigfoot club three times. It was a few minutes before I got a response far away but I did, ha. Once again the knocks got closer than a mile and changed location. They were trying to find me, oh shit. Then from the next tree line over a half mile away I seen a head poke out from behind a tree and look from side to side I seen its shoulders too. That was north of me. I gave another good set of knocks and got a response right away, no vocals just knocks today. Then I started to think about the pine forest behind me and the fact that they like sneaking up on you. I remember this from [downstate location]. Even when you do see them it's always like the first time you met, you always get spooked because they just appear or sneak up, it's just a weird feeling. They stare at you with those wild eyes and neither of you know what to think of each other.

"Anyway, back to today. I knew this young one from the ridge across the way seen me so it was a matter of time. It was only a six- or seven-footer by the looks of it. I kept on knocking at least every five minutes. Now I knew I was in the middle of a family, because I was hearing knocks all around me in at least three or four locations around me. Then I heard a knock maybe shit at least within a hundred yards, with the limb he was using snapping. I knocked right away and just like I figured they were surrounding me from behind so I kept looking over my shoulder once in a while playing along. Then I had to turn around altogether because I heard him break a tree limb off a tree. I studied for a second and there he was, a big boy. You know Chris maybe you're right I might have some magic in me they're drawn to. Well there he was very still now because I had my eyes set on him almost like he was hunting or stalking I should say. He had a look on his face like what the hell does this guy want, it looked kind of pissed off in a studying kind of way. I started making humming sounds, then I surprised him with a whoop. His face changed altogether like oh that's who has been whooping outside of town. Then he looked to the west in the forest, a thick forest I might add. So he was with a buddy, making facial expressions in a calm manner like it's ok he's a friend or something like that. This guy is huge, at first when I glanced his way I thought it was a fallen tree or a stump, then I made out the face. He was crouched up, hunched over. (Solid.) He has a wide black face and fur. A tall forehead and a cone-like dome on top. Big shoulders and huge legs from what I could make out. He was pretty far off, maybe five hundred feet. He was looking around him now a lot. He had no reason to run I was in the middle of nowhere on his turf. So I started eating some peanut butter crackers I packed in my pocket and offering to him. So I ate the whole pack and said mmm every bite, while smooching kisses. Everything understands smooches. I left an open pack in two tree limbs high up for him to get.

"I wasn't that surprised by what happened today, a major milestone. I figure little steps now that they know who I am, I'll be back. The crackers just helped to tell them I'm a good guy.

"So I took a piss and left swaying my arms and looking back every couple of hundred feet waving back. I knew they would be looking or eating the crackers. I feel all giddy inside when I think of it now. Now that I give it a good figure in my recollection, he had to have been five hundred to eight hundred feet away. I should have left more crackers. Well there's always tomorrow."

January 25, 2007: "I went out there at 11:00 am and back to the same spot. Again, it only took about fifteen minutes to get a knock response. I think I found their winter hangout, a nice thick spruce forest."

"Still," I say today, the following summer; it's been hours since we heard that first welcoming knock. "Fifteen minutes can feel like an eternity."

"Two things that keep people from making contact," Jerry instructs me, "are impatience and disbelief.

"Okay, so I started out at the top of the hill whooping my way in. I got there and was knocking. The knocks started back and they weren't just tree trunks expanding with the coldness. A big knock out a half mile followed with one close to my east. The sound of thumping snow pack again around me. I couldn't see Buddy run but I knew it was him by the sound. Branches crunching now and again by mistake or just a spurt of quick movement. But today he wasn't where I seen him last time so that means I wasn't seeing things last time, just checking myself. I looked north and seen a head and shoulders that moved between vertical tree trunks maybe a half mile across the field up the hill."

"Wow," I say, "a half *mile*?"

"Yeah. People don't look far enough, hell they don't even look a *quarter* as far as they should when they're out. It happens a lot. You just have to know what silhouettes to look for and be able to tell the difference between tree trunks and legs, arms and fur. I tell ya, Man, we're dealing with the Abominable Special Forces. These guys are good at what they do.

"So I started humming very nicely. The knocks continued but only north of me after the ones to the east and west within a few hundred feet. Then a while later I heard a hum grumble very low in tone in the woods behind me followed by a big knock to the right of me which was where he was. Big knocks deeper in the woods a half mile to the northeast going deeper and closer. I finally caught on. They wanted me to go deeper in the woods, too, so I did. I went west on this path a half mile out of sight of the other area when I heard a burst of movement from where he was to keep an eye on me. I found a stream and stopped and hung up a bag of apples and crackers and a banana, since I found a peel a while back. Oh yeah he only took one cracker from the other day, you could see he ripped the bag open and dropped it in the snow. Also, I found a set of some tracks with an at least a six-to-seven-foot stride downhill. I had to do a split to make one step.

"I was hearing something behind me, thumping, no reason to worry they're just checking you out. I had the feeling that if they could only draw me out deeper they could attempt to approach me. It felt like at least six individuals around me. One kept knocking and would go deeper, knock, go further knock, and then come all the way back and knock in the same area. I would whoop and get a single knock in return immediately. I was in my chair and looking north towards the food bag knocking my chair leg.

"When I heard something over my shoulder I took a look and out of nowhere a huge knock came from in front of me when I wasn't looking. I instantly responded with a loud (nice) whoop and got out of my chair, and a rock was thrown at a tree a hundred feet in front of me. That means you're not welcome I've heard, but they just don't know me yet. I feel they're testing my reaction, so I stayed friendly. Then there was a vocal from what I thought was a youngster who couldn't hold himself back out of curiosity. Not a big one just a bleep or a boop. It's funny because when I'm home thinking, I get a bit spooked about what could happen with a nine-hundred-pound wildman, but when I get out there that all changes. It's like we're both playing opposite sides of the same game board, both very curious and very unsure."

Jerry points to a branch that forms part of the lean-to opening. "I came back here one day to find a single blade of green grass just hanging there. And this was in *February*."

It's twilight now, and it comes to our attention that we're starving, unwilling to follow our loose plan of fasting all night. Hiking back out to town, though, our dim foresight suddenly turns to luck. A ridge above us, probably two hundred feet away across a green and yellow field, begins to ring out with voices. Just two, but nonhuman, calling back and forth. It's a cross between a howling and the long sustain of great big bells, and since there are no wolves (or, okay, *virtually* no wolves) in Upstate New York, and since this vocalization contains, anyway, zero of the yipping uptake or the group-choral quality of coyotes, nothing but the high song portion itself, and since it comes from the very area where this straightforward young man has claimed to see and hear "them" for months, I feel my belief level shoot from 50% to 85%.

"Captivating, isn't it?" says Jerry.

"I think it's a female and a juvenile, a couple hundred yards apart. That's what it seems like to me. Hope you're getting this," he says, gesturing to my video camera.

To the unaided ear, at the time, able to extract the remarkable from the weave of usual summer sounds, this serenade could not be more plain, and would have made the two-hundred-and-thirty-mile drive worthwhile by itself. But my camera's mic picks up way too much foreground, too much buffeting breeze and trilling finches, so that the soaring background voices in the forest, when I play the footage back the next day, are vanishingly faint.

This acoustical visitation feels generous and primes us, of course, for the night ahead, but after we eat—Jerry's mother fixes us a pasta dinner—and find our flashlit way back down to camp, the show is over. Jerry sits up by himself for a couple hours, hearing nothing, then he listens while I sleep, sedated his mom's Italian sausage. Once, at 4:15, he hears possible footsteps, but not clear enough to wake me.

They must be off tending to other business.

Over breakfast, before I head back home to Vermont, he reminds me of another of his favorite axioms, what he uses to keep himself humble. "They don't need us."

Snowy December, six months later, and I've come to visit Jerry again. We reach his spot in the woods at four in the afternoon, just dusk. Since I was last here, he's constructed an impressive hut out of trees and branches, leaves, mud, and moss. We start a fire and try to get warm; it's sixteen degrees, and we huddle on a log seat, drink coffee from a thermos.

In a while, we take a short walk up a logging road. The snow is a foot deep and the night is perfectly still. Jerry makes soft whoops, and we listen. The best wood knock I've ever heard greets us, and we slap gloves in celebration. I'd estimate it's a hundred and fifty feet away, but given the prime acoustical conditions, it comes to our ears clean and immediate, THWACK, but without much heft behind it. More like THWACK, then, but definitely all caps, unmistakable. It's like hearing a word in deep space, such an affirmation of our being way out here at all.

"Look how smart these guys are," Jerry says. "They just have to make a little knock here and there and they know where everyone is. That's their radar."

Fifteen minutes later, on a second exploratory foray up the hill, we hear another reassuring wooden smack, even closer.

"How come it's such a dainty knock?" I ask.

"That's what they do. They don't want no one to know they're here. Last winter, that's all I heard, 'cause the one would be here, and one would be over there, and they'd be going back and forth. And a light knock like that here and there. 'cause these guys are so smart, they know what to do. They're pros at it."

Back in the hut, Jerry whoops, and not thirty seconds later we hear a series of three knocks from the third-of-a-mile range, quieter than the first but only thanks to distance: obviously more powerful at the source: BOOM-BOOM-[pause]-BOOM!

Delighted, Jerry emphasizes, "That's right after I whooped. And that's right in the swamp, and no one goes back in the swamp. See, that's the kind of knocks I'd get last summer. Three in a row . . . fifteen seconds . . . three in a row."

After an hour of shivering, banking the fire with twigs, we tap free of snow, and we hear a branch snap not fifty feet behind the hut. We're bowled over. It's not possible that a deer, say, has *stepped* on the branch, because of the muffling layer of snow. It's a branch snapped up in pure air, by something with sure hands, as can be readily discerned from the audio recording.

Even though we return at sunrise (twelve degrees!) and do our best to draw them in again, that snap is our closest pass, and of course it makes perfect sense that they'd keep their distance, not show themselves by daylight or approach any nearer at night, given how radically Jerry's breaking his normal routine, feeding a whole new person, an unknown man, into the mix. Men hunt. And two men are, from a strategic point of view, much more than twice as hard to deal with than one man. It becomes a group-on-group situation and injects a major new level of concern—the possibility of being triangulated or flanked.

Back in August, as he entered the area one day, Jerry found himself trailed by the one he calls "Dude," an adolescent male, six or seven feet tall but still dwarfed by "Buddy." Arriving at the hut, Jerry said, "Hey . . . there . . . Dude" as he slowly brought the camera up to his eye and snapped a shot. It's an extremely low-resolution image, but it does show the figure leaning out from behind a tree at approximately two hundred and fifty feet. It's the sort of picture that wouldn't convince anyone who hadn't come to know, and finally trust, the context of these encounters.

All during our mini-expedition, which is entirely satisfying if not earth-shaking, I am suffused also with the experiences he has conveyed through his emails, this past summer and autumn.

July 8, 2007: "I had a great time out there today. I whooped in the field on my way in. As soon as I got in the woods I was being followed. It's hard to explain. We were making our ways to our normal spots where we check each other out. When I would stop walking he would too. When I got to my hut I could see someone was playing out there. Well I heard him getting damn close behind me so I looked and there he was concentrating on his next move, ha. I caught him. He looked right at me right away. Shit man it's been about eight or nine months since we gazed into each other's eyes directly since we met. And he was fuckin' close, like three hundred feet away no more than that. And it was Buddy himself crouched down low like when we met. I got a good look at his feet and can see how he leaves twenty-eight-inch tracks. He is *huge*. He has a gorilla head just a lot bigger than one from Africa.

We stared each other down. I could hear what he was thinking: Oh shit I got too close. With wide eyes and an open mouth he was dumfounded. So I started to talk like a baby real sweet-like ya know. So he wouldn't feel scared to be that close. I turned to take a pee and out of the corner of my eye he took off in a rush so he's still very nervous of me. I can respect that big-time, this will take time like I thought. When he took off he took a few steps and looked to see what I was up to and before ya know it he was in full stride and gone out of sight quick. Man they can blend in good out there.

September 14, 2007: "Today though there was an honest effort on his behalf to make contact. I was surrounded by like six of them within a half mile, all around me, with caveman talk and a nice three-minute-long set of knocks some light vocals behind me to mimic my babytalk, that was Moma I think. And running through the woods.

"It happened, they finally made the next step to friendship. Buddy was happy, I could feel it, a feeling of excitement all around me today. Maybe there's a new baby. I knew who I was looking at today right away. One look and it was Buddy. Huge! He's done this to me a lot, follow me in like that. If you recall I've brought this scenario up before. But we've made a good step today. They might want to speed this up a bit before winter for their own reasons, Hey. So I'll know what kind of eats they like, right? I think the quick change in weather has brought this new approach on maybe. Once they figure out what a PBJ is we're set. I'll bring one out next time

"He left a gift today, a walking stick in the trail like he's done for me before. The hut was tampered with but not messed up, like a passing-through inspection. The candle was moved to the other side of the rock and a few other things were moved too. I could tell it was him, Bud, because it was like he flipped it around with his fingers.

"I also thought I heard a baby squatch out there today with Mama. It didn't cry just a whimper once in a while, I was right about my hunch of the baby. That's probably why they were happy today they probably just started bringing out the baby. That's why they were on that secluded hill a mile away. They need food. I'll bring a food gift next time. For Mama. I'll have to make a separate area instead of my hut but in view of it for the food to be offered at. Soon enough the leaves will be gone.

October 25, 2008: "I've been getting little whaah's from the new addition, my friend, and Mama's been trying to shush up this one I can hear. I believe I did as good as I did today because I didn't have my camera, it wasn't charged in time before I left. I played some METALLICA for them. They liked it a lot. I guess that's what I'll name the baby, METALLICA. They really can do good human vocals. It was like a celebration in the woods."

Family D: Texas #1

This is a family of six. The parents are in their forties, with two grown children and, still living at home, a thirteen-year-old son and their daughter Rachel, who is fifteen.

Things have been happening here for twenty-six years. I think the first episode I remember is being out late one night (about 10:30), feeding some dogs I had in a pen out behind the house. I was by myself and it was dark. I believe the weather was a little on the cool side so maybe it was in the fall. Anyway, as I started to open the gate to give the dogs food, I heard a really low growl and then teeth clicking together. I've always lived in the country, so really I'm not afraid of being outside after dark by myself. In fact we grew up playing outside all the time till the early hours of the morning.

But these teeth sounds were close and very loud. It scared me. It scared me a lot! I didn't recognize the sounds and had never heard them before, so I just dropped the food and walked very quickly to the house.

Over the years we've had some things disappear as well as heard strange sounds in the woods. We never really thought much about it, though. Until Rachel came in one day saying she had seen a bear. As she described what she saw, I was thinking it sounded like a Bigfoot but was a little skeptical. I'd never really believed one way or another. Then when I talked to my oldest son, he mentioned he had just seen one walk in front of his truck on his way home late one night.

That's when we started searching out sightings on the internet and found the BFRO. It's hard to say how frequently things have happened, since we weren't really looking or paying attention to such things. Since we've been watching for signs, we notice things almost daily.

Mostly it's just calls in the woods, but sometimes there are close encounters. They usually show themselves to the younger kids, more than to us grownups. Our daughter is the main one to see them. She's actually seen the mom and twin babies!

If our chronology is accurate, we saw the mom pregnant outside our backdoor early last December. She was huge and we think her giving birth was imminent. Then Rachel saw two babies in our pasture that ran for their momma and they were about the size of a mid-size dog. They ran on all fours, like a chimp runs.

In terms of "gifts," yes we do leave things out for them sometimes. I wouldn't say we do it regularly. Sometimes when Roy [local BFRO Investigator] is here, he brings food for them. We especially leave food for Christmas. They seem to like it. Usually fruit, pancakes and hot dogs.

Rachel has show rabbits. One time, one went missing and a big rock was found in its cage. So, yes they have left things. What's really funny though is Rachel had several rabbits die pretty close together (they die pretty easily), and she got tired of burying them (because the dogs kept digging them back up). So, so left several right in the edge of the woods behind the house. The next morning we found them on our front porch. I think that was a hint, they didn't want them on their front porch.

Roy leaves a bag hanging in the tree over a little ways from our house. The other day they started calling to us and put the bag in the tree right behind our house . . . like they wanted us to fill it up again.

Rachel said the mom she saw was beautiful. She had very pretty features. Fine features and very feminine.

But she was "zapped" that one day when she was getting close to the babies. The whole time she felt paralyzed, the mom was cooing to her. She said she wanted to respond and talk to her, but was unable to move or say anything. Then when the babies got out of sight the mom turned to walk away. Rachel followed and called to her asking her to please not go. She turned to my daughter, looked at her, and then walked into the corn. (The corn field was full grown at that point in time.)

I sometimes feel like they are better off than we are because it's so simple. They just live to be together. They don't get caught up in the day-to-day struggle of materialistic things we get hung up on.

Are there stick structures around here? Oh yes. In fact, there's basically a village created out of brush and twigs. It's amazing. It has several huts that are very clean. It is our impression that they travel around from place to place within their territory, never staying too long at any one location. From the calls we know that they are all around us.

I hate the groups that want to kill one to prove their existence. What if that was a member of their own family? How would they feel then. They obviously care about one another. They also live in family groups with a mom and dad.

Rachel is the main one to talk to. They seem to actually interact with her. Usually when she's alone out there with the horses. They really like her. I'll let you talk to her. As long as you don't reveal our specifics, it's fine. It's really not so much a matter of our personal privacy as it is *their* privacy. I would feel horrible if something happened to them because I started this whole thing with the BFRO. We just didn't know much about them in the beginning and now we feel like they are neighbors that are friendly, but like their privacy. They seem to be so protective of one another and truthfully, that includes Rachel. They seem to watch over her. In the beginning we kind of feared they might take her as a pet, but now we know they would never hurt her.

It's not strange to any of us, to me, my husband, my daughter or my son. All of us have seen them now. We just wish we could get some good video. We've got three cameras up, but honestly, haven't gotten much. Roy got some good thermal views last year of three watching me and Rachel. It was on a Sunday about two in the afternoon! They are here all times of the day and night with no rhyme or reason for the timing. I mean you can't say, "Well, every Sunday they'll be here," etc. They just show up when they show up. It's just that it's pretty often. I think they feel safe with us because we interact with them through calls and talking. Then we go and leave them alone.

Even a couple of our neighbors know of them and it's common conversation about the big hairy naked people. But they don't seek out encounters and are not outside that much so don't really interact with them. However, the Feet walk across their roof as well. They are always walking on the roof here. Pretty funny. Last night someone was banging on the outside of my bedroom. I didn't get up. But it was pretty loud. They knock on the wall, the walk on the roof, I think they may have even jumped on our trampoline before. One night I heard them walking on the roof and then a stumble and bang bang bang (like they fell off the roof) and I went running out to check on them and they were already gone.

We live in a junkyard basically and I think it makes the Feet feel safe. They come around all the time. In fact Roy brought someone here about a week ago and they went walking in the woods about this time of day and walked up on a napping adult. It ran off, startled the humans. Pretty funny. Of course, he didn't take his camera, 'cause he didn't think they'd be out. I keep saying we need to hook up Rachel with a hidden camera and microphone all the time.

From Rachel

I show rabbits. One day I had a show, so I had to get up about 4:30 AM or so and had to go outside to get the rabbits. I had a headlight type thing on my head when I walked outside, the dogs were going crazy and when I looked into the pasture there were four eyes looking back at me, big eyes. I couldn't figure out what it was so I got closer and closer then my dog barked and they both got up and zipped off into the pasture. I shined the light over there, and there were two things laying there. I thought it was maybe a big cat, I couldn't tell, they had BIG eyes. They were looking all around and looking right at me. I was shining the light on them and they were reflecting back. So I started to walk towards them, and when I did, Zeke nipped one of them, and they both tore off running really fast, so fast I don't even remember much what happened. They went straight up under our neighbor's fence and stood upright and went into the woods, I was like, Oh, that's what those are. It was two baby Bigfoots. I was shocked!

And I shined the light back on the cornfield, and I could see them running through the corn, and I was like, I'm going to go get a closer look, so I ran up to the fence, and I get within about six or eight feet of the fence, and Mama walks out, puts her hands on the fence and just looks at me. I'd seen her a couple times before. She starts making this humming noise? And I couldn't move, I was just like stuck there. I was so scared, couldn't make a noise, couldn't do anything. I was getting really scared, and then she started making like this cooing noise, and then I wasn't so scared anymore. I just kind of stared at her for a little bit. And then, she looked at me and turned around and went back in. I said to her, "Wait." She looked back at me briefly but kept going. I still went ahead and got my rabbits ready and left.

She had a very human face, very pretty. She didn't have hair on her face either. Black skin, chocolate black. More like charcoal.

But way before that, I was the first in my family to say, "Hey, there are Foots in the woods," and of course no one believed it until my big brother saw one. They have been there all my life, I just didn't know it.

I was walking over to my grandmother's house the first time I saw one, and it was probably about dusk. It came out of the woods from behind the little house and walked upright towards our house. And then it got on all fours and ran back in. And I got all freaked out and went and told my mom, "There's a red bear," and she wouldn't believe anything about it. That was when I was nine or ten.

So, I got really upset because she wouldn't believe me. Then the next time, me and my cousin were outside playing in the dirt, because that's what we do, and something came stomping through the woods, breaking branches, making all kinds of racket, and we saw something big and reddish brown-black, running through over by the little house. We got really freaked out and ran inside, and were like, "The bear's back." And she still wouldn't believe us.

And so then, like a week later, my big brother was coming down [route name] right over there, and something ran out in front of the car, and he told my mom, "It looked like a bear but I think it was Bigfoot." And then she put two and two together . . . and *then* she believed it.

My brother feels as close to them as I do.

I feel like they are part of my family. They are always there.

In the beginning, it probably took six or eight months, and then I started messing with them. I used to feed them almost every day. I made this . . . I just put some flour, eggs, and milk and sugar and syrup, and put some sprinkles on it and cooked it, and would put it out there and come back a couple hours later and it would be gone. And I'd have it up so high that nothing else could possibly have gotten to it. Sometimes off the eaves of the little house. Or sometimes I had this huge table that was taller than me, I had to get a ladder and put it up there. It would be gone every time.

I stopped putting food out there, I got too lazy.

Sometimes I'd just look at them and they'd run away, but after a while, I'd look at them and they'd just kind of duck down and stay there. If I talked to them, they wouldn't really say anything back, they'd just kind of go lower. If I got closer they'd run away. But if it was after dark you could see their eyes.

They leave me stones sometimes, and they whistle at me all the time. And they imitate my mom and say "Rachel" all the time. They will act like they are my dogs crying far away and as soon I get far far away looking for the dog they say "Rachel" back at the house, my mom's voice. Then when I get to the house my mom's not home, then they make the dog noise again, making me run back over and over till I give up.

Just the other night, someone was mimicking my little brother calling for the dog. We hear that a lot.

Or they take stuff and hide it and when I go to get it they try to scare me but it doesn't work.

The big male isn't friendly but he is only around sometimes and yes he does do the [infrasound] noise thing to me. He does not like me anywhere off the yard. He sometimes throws rocks beside me but not hitting me though.

Back to the Mother

Lots of activity recently. The other night one actually waved to us. It was midnight, but bright moon. It was probably one hundred yards away. We've watched each other from that close before and even whistled or called to each other, but that's the first time they actually waved back. We were pretty excited. We try not to push any situation to a point where they feel uncomfortable. So after we called to them and waved a couple of times, we went in. We're trying to show them that we know they're there and we aren't trying to hurt them. We're non-confrontational. I think that's why they came to our house when the helicopters were pursuing them in the woods that night.

We've been having helicopters flying low over the woods at night, usually late, and sometimes people flushing them towards our house from the back of the woods. We've found lots of evidence that someone has been trying to find them.

As soon as we can cough up some extra money, we're going to put up more cameras and get another DVR. We have one now with three cameras up. But, frankly I haven't had time to watch it lately. I need to try to get that done.

As I mentioned, Roy got three on thermal camera last year. It was really clear. There were two big ones and one baby crawling through the woods towards our horse barn. It was 2:00 pm and Rachel and I were out there taking care of the horses. You can see two larger figures crawling on their hands and knees and then you see a little one crawl along behind. You can make out their lower legs and feet. Rachel and I didn't even see or hear them. It just reiterates how much they watch us.

BFRO Investigator Roy writes, "We made several attempts to make sure it was 'Squatches on all fours, and it is. We went back there and took turns trying to mimic them and didn't even come close to the same size."

Just last night, I was watching Rachel run to Grandma's, when I saw one at the "little house." It was sort of pacing and watching me. They often hang around the little house and watch us go back and forth to Grandma's.

My husband's grandmother lives next door. We live in the country so next door is about a hundred and fifty yards from us. The little house is also about one hundred and fifty yards southeast of us. It's pretty much just a separate bedroom set away from Grandma's house. She got mad at Grandpa about something one time and made him build it for her as a retreat. Anyway, after Grandpa died, Grandma has been too scared to stay alone. Rachel and/or her brother have been spending the night there most every night since that time. The kids wait until as late as possible to go. We pretty much have a timeline of 10-12 pm when one of them runs across. I usually either walk them or watch them from the fence that runs in between our two properties.

The Feet usually hang out around the little house for the nightly "show" of us running across and playing in the pasture. We often see one run around the little house when we walk out to the fence. Sometimes they call, but most of the time they just watch.

Grandma has heard things and thinks people knock on her house and walk on her roof. She does not believe it is a Foot. We actually spoke of it to her once, but she didn't believe it. Her memory is not that great, so she doesn't remember any of us talking about it. We don't bring it up.

I get excited any time I see them; however, we do see them a lot. My younger son really likes to see them. He comes running in to get me. For me, it's always at night. During daylight hours, they only show themselves to the kids.

Family E: Texas #2

This site is one hundred and two miles from that of Texas #1, but the two families have never met, or communicated, as of the time of these testimonials. The speaker, fifty-four, was chemically poisoned at work in 1993, and has since been on disability. She currently lives alone in a "flimsy lap-siding house," though in the 1990s she lived here with a husband and four teenaged daughters. Before she moved back in late 2007, the house had stood vacant for five years. Her property and the few properties nearby are surrounded by hundreds of acres of forest.

The late 1990s

When Bill had the rock fight, we never could figure out who he'd had the rock fight *with.* so we just kind of dismissed it, and after a period of time it just goes into the non-thinking part of the brain. This was in 1998. It was dark out there, there was really no moon, and that area is covered by trees. What he saw was built much like my little spindly daughter. I think it started out as he thought maybe she just chunked a rock at him and inadvertently hit him, but it pissed him off. And so he just reached down and grabbed up a rock and flung one back at her, and hit her, what he thought was her. It made a sound like someone getting their wind knocked out.

Well then they got into this rock fight, and he said it was quick, very agile, in the dark, which he didn't really understand how she could see where she was going because it was so damn dark. And being on the run and side-arming rocks and just beaning him time after time. I was able to look at him later and there were like nine spots on him, because when he'd see the arm move then he'd turn his back to it and it'd get him right in the middle of his back. He must've had eight or nine big ole knots in his back, and a couple on the back of his head, and one on his forehead.

Well, he came inside and he was loaded for bear. He was waiting for her to come in and I said, "What's going on?" and he said, "I'm waiting . . . just never mind." Then I said, "Well, what's going on?" "I'm waiting for Allison." I said, "Allison," and she comes in from the bedroom, and no of course she wasn't dressed all in black like Ninja Child.

At first he started shaking and then he turned white as a sheet, and I thought it was from getting beaned in the head, like he was going into shock. I thought he'd run into a tree, because he didn't tell me he'd gotten in a rock fight right away, all I knew was he had a big ole knot on his head. But then when I started looking at him, you know, you could see a knot on the back of his head, too. It wasn't till after the kids settled down for the night and we were laying in bed, and I said, "You wanna talk about it?" And he said, "Not really." And I said, "Bill, *what's* going on?" And he said, "Allison never left the bedroom?" And I said, "Bill, she's been working on that school paper all evening." And he said, "I got in a rock fight." And I said, "With who?" And he said, "Well, I thought it was with Allison." I said, "Well, who was it?" And he said, "I don't know, and I don't wanna talk about it anymore. I finally got tired of the rock-throwing, I was just gonna chase her down and whoop her butt." And he chased her down into the ravine and he couldn't figure out how she got down there so dang fast. And where she went.

"I think they've been here all along. You know, with three teenage girls, and two of them crawling in and out windows and smoking marijuana, and sleeping with boys (those were *his* two), I was trying to hold down the fort with that, and I was still recovering from being chemical poisoned, and then we had forty heads of milk goat and sixty chickens and thirteen hogs that we were raising from babies and, you know, there was plenty to keep my mind going and my body tired.

"So there were a lot of things I would dismiss. Things being moved outside. I'd leave a hoe right there leaning against a tree near the garden. I'd go out there the next day to finish up and it's not there. And I'd find it out by the goat house hanging in a tree. And I'd think, Haven't y'all got something else to do besides mess with my tools? I wish you girls would just leave shit alone, and so, you know, there was ongoing confusion here. All the time. I would get my flower pots and stuff . . . you know, when you have gardens you've always got *stuff.* And I put all my flower pots in one area and I'd go back and half of them were gone or moved. Mostly I'd notice it overnight.

Then there was the voice trick. The girls would be at school and I would hear, "Mom!" from the woods. And I'd think, That couldn't be the girls because they're at school. But often, you know, it would be so real I would go and check just to see if maybe they got a ride home from school because they were sick? But we had to sign them out . . .

Or they'd be at home and come inside: "What do you want?" And I'd say, "What are you talking about?" So you know, these Forest People were imitating me and imitating them. I think they'd just sit up in the trees or whatever and that was their entertainment. *Watch this one, watch this one.* Like a prank.

Back in 1999, my daughter had a teenage girlfriend from Arizona visiting. Allison and Michelle (the guest) played in the woods for hours on end, making dams in the creek, exploring, climbing trees, etc. One evening, just before sunset, they came busting out of the woods running as fast as they could go. I could tell they had been frightened, but they ran right on past those of us sitting on lawn chairs and hid in the bedroom in the closet. I went in and asked what had happened. Allison told me they got scared and came home . . . but told me no more about *what* had scared them. I believe it took them a couple of hours to finally open up and tell us that they saw a large bear-type creature come down feet-first out of a tree, land behind a bush, stand up on two feet, then side-step behind a large tree trunk and peek out at them. Allison said that the creature didn't move like a bear, nor did it have any "ears" like one. She said she and Michelle had run out of the woods following the well-used path, and the creature followed them, keeping pace—but through the woods to the side of the path. Once they reached the mowed yard area, the creature stayed behind. The girls refused to go into the woods after that, and wouldn't remain outdoors when the sun started setting. After that time, we teased my daughter about her "big, hairy friend" in the woods. Since none of the rest of us had experienced anything remotely similar, she became the butt of some pretty mean jokes.

From Fall, 2007

The house had stood vacant for five years when I decided to move back by myself. During this past summer, I came up more and more often and it must have been about the beginning of October I came up and somebody'd taken a big *crap* in the middle of the living room floor. It was just disgusting. I'm thinking, What has this person been eating? I cleaned it up, scrubbed it with Lysol, scrubbed it with bleach, you know, and the smell still was in here for four days. That's how pungent it was.

Then I went into the bedroom and it looked like somebody had brought in a big section of rolled hay. You know how when they bail rolled hay it's in layers? This was one layer. It was about a foot thick by five foot wide and about eight foot long, and it stank hideously, and I thought people around here are using this as a flop house. But this is what I thought was really bizarre, there was a pile of red surveyors' ribbons and orange surveyors' ribbons and different-colored Christmas ribbons and strings and little pieces of wrapping paper.

When I moved back here, there were twelve windows broken out of the house. Two of them looked like somebody had jumped through them from the inside. The back windows looked like they'd been Kung Fu kicked *out.* And I thought, You know what, that's a lot of wasted energy, you kids just have too much energy. Prior to October, some of the windowpanes were broken but the glass was inside the house. But this glass was broken from the inside and pushed out. In hindsight, it kind of scares me: They know the lay-out of the house, too.

One of the ladies on the habituators' forum said, "You know what, I bet you they decided this was a good place for them to get out of the weather, and they are rather warm-natured . . . they probably kicked it out for air circulation."

Because the house sat empty for so long they may have thought that I'd left it for them. So there was like a failure to communicate.

I had my first sighting on March 2, 2008, about 9:00 am when the dogs started pitching a fit. It pretty well changed my world as I knew it. It's one thing to *hear* about the Bigfoot, it's a whole new world when you actually see one. I got probably a twenty-second viewing of my hairy friend, which was quite enough for a first encounter. I was still shaking two days later when the researchers showed up to take my statement. And, when they found the knuckle print, I had to sit down. It was sort of unrealistic until that point. Once they got the two-by-four-by-eight-inch board out and I found out *exactly* how big this guy was, that cinched it—I was ready to put out the "For Sale" sign in the yard. But, I started thinking back about all the times they *could* have harmed us, and obviously didn't—and thought also about some backwoods rednecks moving in here and causing them harm. Well, I just couldn't do it. Both my daughters think I'm absolutely out of my mind to live here by myself, but I don't see either one of them volunteering to stay here with me!

So here's how it went that day. I normally get up rather early, put on coffee and let my small dogs and one coonhound out to relieve their bladders. This morning, being rather cool, I didn't stay out with the dogs, but went back into the house. My daughter was visiting and was still asleep. The dogs began barking. In an effort to quiet them, I stepped out onto the porch. The Chihuahua and sheltie/rat terrier were in a small pen, the coonhound on a tether. The coonhound, instead of going to the end of her tether, was only about 1/4 of the way extended. The small dogs were penned, but were all looking southward, toward the woods. The dogs had a strange bark, not like seeing a person on the road, or a deer—those barks are familiar. The only way I can describe the barking is that it was rather whiney. I looked in the direction the dogs were watching and was shocked to see a large form. The young growth of pine trees was about three to four foot tall. This form appeared to be twice the height of the pine trees. It was really bulky, with no neck, and blended in with the dappling of the larger trees. It was covered in hair, and the length of the hair on its body was maybe six to eight inches. The hair on its head was slightly shorter, roughly four to six inches. You could call the hair color calico: a mixture of charcoal, gray and brown. Overall, the hair was really messy-looking. I think the shoulders were probably between three and three-and-a-half feet wide.

So I kept looking and questioning myself, and realized I truly was seeing something unusual. I pounded on the side of the house, awakening my daughter to also witness this incident. The large form did not turn right or left, but seemed to be moving slowly but steadily backwards into the cover of the woods. By the time my daughter got outside, the creature had blended into the woods.

Then five days later, she had her own sighting. At 8:00 pm, she decided to drive into town. The house has a long driveway that runs right in front of the pine saplings where I saw the figure. Allison got into her car and drove down the driveway. At the end of the driveway, her headlight beams lit up what she first thought to be a large tree, or stump, in the middle of the saplings. But then this stump started gently swaying back and forth. She turned on her high beams and saw what she could only describe looking "like the lion's head from the Wizard of Oz."

She called me on her cell phone and backed up her car, trying to get the headlights into position so she could see the thing better. Doing this, she took her eyes off it while shifting into reverse, and lost sight of it. Allison figures the incident lasted about five seconds. She couldn't see the torso or shoulders, just a head. She said the face was like "earthy colors," marbled black and dark brown.

There appeared to be a tuft of hair on the nose, but no hair on the cheeks.

She was too scared to get out of her car or drive away, she was basically frozen. After I and my other daughter came out and calmed her down, she finally left.

The researchers arrived the next morning, and here's what they wrote about what they did and found.

"With the witness standing on her front porch, and with an investigator standing at the sighting location with a pre-marked pole, we were able to determine, based on the witness's recollection, that the subject was approximately eight feet tall.

"As is typical of the ground surrounding a growth of young pine trees in East Texas, the sighting area was covered with a thick and heavy layer of dead grass; and the surrounding floor of the mature pines had a thick layer of pine straw and leaves. We performed a detailed search for trace evidence and after quite some time found a fresh impression on a small gopher mound within the pine saplings that exhibited clear digital impressions. Upon examination, it appeared that the impressions represented a knuckled imprint of a large hand, approximately six and one-half inches in width. A cast was made which verified that there were five digits in the print. The print appeared to have been made within the last twenty-four hours as the area had received rain and snow two days earlier. It is possible that the print correlates with the younger daughter's visual incident."

That was seven weeks ago. Since then, all kinds of other stuff has been going on. I had the house bumped. I have a twelve-year-old coonhound, and something was bothering her because she woke me up out of a sound sleep. She was in the living room, just doing this nasal whistling, and I thought, What in the world is bothering that dog? And about that time something hit the southwest corner of my house so hard the windows vibrated. The only thing I could think of is some sort of livestock got loose, or the house fell off the foundation block, or something's trying to get in. So I got my gun, got my flashlight . . . because it made me *mad*, you know? So I put the flashlight under my arm, got the pistol, and the dog and I went outside. Nothing was there when I shined my flashlight around, but there's this little wiry flowerbed border that was like six foot and it was all torn up where something looked like it had tripped in it.

Often, right outside my window at night I'll have this loud "AAAA!" sound, like the "a" in cat. It was like they were trying to see how far they could push me. So, of course the ladies on the habituators' forum said, "Just start talking to 'em, when you go outside working in your yard." So I had some tomato plants and I was out there and talking to them, saying, you know, "I'm putting my tomatoes in here . . . It's been a long time since I've been here and had this garden going . . ." Just jabbering, you know. When I got done doing that I went and sat down on a glider rocker I had put down right at the end of the driveway with the back of it towards the house. And I sat out there for about fifteen, twenty minutes and then said, "Well, if y'all aren't going to talk to me, I'm just gonna go in the house." I got on the second step of the porch and it sounded like I was transported to the Dallas Zoo, in the primate section. It started out with two "Woo Hoo Hoo Hoo!"—two of them doing that. And then it went to that screeching monkeys do when something's been taken away from them.

I started looking around, and north of my house there were five large shapes moving in the trees. They were from about four-and-a-half foot up to I think eight, judging from the trees they were standing behind. There was a row of four-foot trees in front of them and much taller cedars behind them. That's when they started bird-whistling. And then they were frogs. I was getting these calls from just this one little section of the woods. And it blew my mind. So I just stood out there talking to them and all of a sudden . . . I'm still in the process of mowing down because the house was vacant for five years. There was some grass that was probably two foot high, and through this grass I can see something about three foot wide commando style coming towards me in the grass, belly-crawling. All I could think of to say was what I said: "Oh for Heaven sakes, I *see* you." And then it froze. And then it crawled backwards. *Backwards.* And I thought, Oh, that is too creepy.

Meanwhile, they were still whistling and pitching a fit out there across the street in the trees north of my house. And then from *south* of my house there came a bird-whistling so loud it made your eardrums vibrate, so you knew it was no bird on the face of the earth.

A *piercing* whistle, and I looked up and there was one pine tree, about forty foot tall, that was like flapping back and forth, swaying back and forth and the swing was getting bigger and bigger. And about ten foot from the top of it was this wadded-up furry creature. It did not look like a raccoon, and I didn't see a tail. But I cannot honestly say it was a baby Sasquatch. But I'm looking at the tree and thinking, How in the world is that tree getting more and more momentum when the object at the top of it wasn't moving a muscle? Something was moving it from the ground, and the only thing I can figure is it was a baby that had gotten out there where it was too visible, and they wanted to get it back in. It was surreal, and you think, Did I really see that? And I thought, I'm losing my marbles, I am becoming delusional.

When I saw that one up in the tree, it looked like one of those (I think they're called) "burls," a big ole knot that comes out the side of a tree. That's what I thought I was looking at. I didn't know they could climb trees, see that's how ignorant I am of what they're capable of doing. So I went back there later, and where the big knot on the tree had been there was nothing but sky.

May 14, 2008

Recently, my cat DeeDee that I raised from newborn got feline distemper and I ended up having to put her down. Brought her home, wrapped in a towel, put her in a box, carried her out and buried her. The next day, I was in the backyard mowing and I hear my cat call from behind the goat house. DeeDee had a distinct meow, I knew that cat anywhere. And I thought, You sorry bastard, you don't know she's dead. Then I thought, Okay, I'll play along with this, and I went, "DeeDee? DeeDee Kitty," and it did it twice more. Later on, I started thinking that maybe it was making this sound to comfort me, you know like when you see a photograph of someone who's gone?

The other day, I was walking out behind the goat house, in the middle of the afternoon, and I heard from the woods what I thought was a frog at first, except it said very clearly, "No bite. No bite." Really it was like a cross between a frog voice and a person's voice.

Then I suddenly realized, and it made me laugh. See, I'm still trying to follow the advice of the ladies on the habituators' forum, so I've been going out to the edge of the yard and talking to the woods, saying like, "I'm just an old woman living here by myself, I'm not gonna hurt you. I won't bite. See, I don't even have my teeth in." When I was chemical poisoned all the minerals got leached out and my teeth started breaking off below the gum line. So I just went ahead and had them all pulled and got dentures. Now, with all this stress here at the house recently, I've been clamping my jaws so much my gums've gotten bruised. So I don't wear my teeth as much.

Now I've heard this a few other times, too: "No bite. No bite," and I almost feel like they've *named* me it.

June 11, 2008

Last night, I had to let the coonhound LuLu out to go to the bathroom. She's on a forty-foot rope. And I'm getting ready to go to bed and went out and called to her and she wouldn't come in. And I called her again and she wouldn't come in. So I went off my porch to the south, fifteen foot, and just as I came around the edge of the house I hear this "HUH!" And it was so loud I nearly wet myself. And I said, "Look, I'm just going to get my coonhound and I'm going right back in the house." And I could hear it breathing. It was standing in the shadows. See, the only light I've got here right now is the front porch light, which does not shine on the south side of the house, where it's completely, pitch-black dark. I could hear it breathing as I was shaking, grabbing the rope and bringing Lu in, but Lu was almost like she was frozen. And I don't know if I drug, carried, or how I got her in the house.

It traumatized me. They are bound to know that this type of action instills fear. I go out there and I'm telling them, you know, "I live here by myself, now when you make the loud noises close to my house it scares me. I feel like I'm being *threatened*. When I heard that breathing, all the hair on my body stood up, and my heart was beating so loud in my ears. Had it approached me any closer I think I would have had a coronary. I said, "I think I'm going to go in the house now." And of course I know my voice was just shaking.

June 16, 2008

Today when I was mowing, mowed the front yard, didn't have a bit of problem till I got on the north side, where I'm trying to push back the growth from the woods and reclaim my yard. And where that one was commando-style coming up through the grass?

When I started mowing through there, my heart started pounding. It wasn't because I was thinking about the Ninja guy, no I'm thinking about are there any rocks in here . . . . All of a sudden I had this unexplainable fear, and I yelled out, "You bastards, you are *not* taking over my property!" And my heart's still pounding, and so I tilted the lawnmower up and ran it into that grass, and whatever it was just *left*. So then I'm going in the back, and I haven't mowed way back by the goat house yet, I'm getting there, but the whole time I am mowing I get hit with like "Essence of Gym Locker #3," and "Ode to Skunk," and the third time smelled like dead body. And after the third time I said, "Look, I know you've got all these acres of woods, all I'm asking for is my *yard*." And I said, "Whether you like it or not, I'm mowing." And the smells quit.

These things . . . they are agile, they can be vicious . . . you know, any primate pushed can be vicious. I wish I could get you down here and put you over on the other side of the house and let you experience this. I feel like I'm under siege at night. While I was mowing I was getting assaulted with these odors and my heart's pounding because I know they're close, and I started thinking, What is the *purpose* of them making themselves known. There's bound to be some sort of reason. There's nothing I have that they could want, except knowledge. It's not like they want to live in my house. It's not like they want to borrow my *truck*. If they ever got to know human beings and what we are capable of . . . . I feel like a guinea pig. *Okay, this makes her anxious, and this makes her feel good, watch her relax*. What would happen if they decided, *You know what, we want to take back this area*? There's really no way to explain it except I almost feel like I'm in a war zone. At night, every single window is covered.

July 11, 2008, in a Telephone Conversation After Midnight

—I can just barely make him out.

—You mean hearing him or seeing him?

—I'm seeing him. He's hiding his face behind a tree, but he's about two foot on each side of this one-foot tree.

—Is he moving at all, or stock still?

—He's just swaying, with his head right behind the tree. Oh, that is so strange. He's just standing out there behind that tree.

—How far away?

—Forty foot?

[She comes back indoors]

—I'm still shaking. Had you not been on the phone, I would not have gone out there.

—Can you peek out any windows, or are they all just completely blacked out?

—Let me go into the bedroom. I got sheers on there that I can hide behind. I don't see him behind that tree, but that doesn't mean he hasn't moved to another one. Hold on just a second . . . Okay, I have a half-moon-shaped driveway, and he has moved to the other side of the driveway . . .

—What's he doing right now?

—They just stand and *look.*

—Behind anything or—

—Behind another tree, but he's not really all that behind it now, it's like half of him is behind it and half of him is not, because I can see where his head is . . .

—What's the light source? Is it moon or stars?

—The only light I got right now is from the front porch and I can just barely make him out, only because he's swaying. I don't understand why they sway. If they'd hold their asses still they wouldn't be seen so easily.

—I know, but often people say they sway, like gorillas. As though they need to be more intimidating than they already are.

—I'm wondering if it helps their binocular vision.

—That's an idea, yeah.

—He's walking off, he's going toward the south, he's going back over to the pine grove.

—You can see him walking?

—Yeah. It just looks like a leisurely walk, just woop-de-doop, you know.

—Aren't you just amazed that they even exist?

—It's sort of like looking at a gruesome car wreck. You can't take your eyes off of it, but you can't dare look away. I feel intimidated more than anything else. I'm amazing at how graceful they are. When they move out of view it's a glide. You don't see them *step*. It's like they're on tracks, and it blows me away because you can't hear the bastards. They are big, they are hairy, they're everything that nightmares are made of, but they're not imagination any—Eye-glow, hold on. This one's shorter. All I saw was the eye-glow, in the same area across my driveway. I'm shaking so bad. If I drop the phone I'll grab ya as quick as I can. I don't understand why they're so damn curious. From watching me the months I was here, without anybody here, from Halloween until January 10th, until my daughter moved here. So for all those days, I was here by myself. So what is so damn *fascinating*? I got an epiphany today where it was like, We're losing our spot on the food chain, folks.

—We're probably like a car wreck to them, too. If they have an opportunity, where they don't feel threatened, like in this case, and they can just feast their eyes on a human being in its—

—In its own little habitat, too?

—You're like a zoo to them. If they'd wanted to hurt you, they'd have done it a long time ago.

—Unless they were sizing me up.

—They're probably just playing around with you.

—Well, I *don't like the way they play*. Oh there goes another one. That one was little, though.

—What's he doing?

—He just turbo-charged across the driveway on all fours, and disappeared into the trees. Man, he was *quick.*

July 19, 2008

I've made an all-out effort to repel them.

First, I did a big hunt through the house and scrounged up four cameras, which don't work anymore. I mounted them on all four sides of my house.

Second, I spread chemical crystals, flea and tick repellent, on lots of fireant mounds and by the edge of my lawn. The ladies on the forum suggested this, because it's worked for them.

And then third, I went around my property line and told them *again* that this is *my place* and that they could have all the miles of woods around here but to leave me my house and yard.

Late yesterday evening I was able to sit on the front porch (porch light on) without heightened anxiety. The coonhound, although alert, was more relaxed. She did not bark or whine even during the period of time she was alone outside, which has not happened for many weeks. Lu was so relaxed while I was outdoors she laid down on her side and even closed her eyes.

Whether the result of walking the property and establishing "boundaries," chemically treating the mounds of fireants in the yard, mounting dummy cameras on the house, or even the beneficial psychological effects of me actually doing something constructive—I slept like a baby last night! Nothing hit the window screens, no vocalizations in the yard, even the dogs and cats appear to be much more relaxed.

Right now, I get the feeling the Ancient Ones (as I like to call them) are a *bit* upset with me. When I talk to the woods, it almost feels I am talking to myself. No longer do they give me a responding wood knock. I feel a bit guilty, but I certainly did enjoy that full night of sleep last night. And I definitely enjoy being able to sit on my front porch without being intimidated.

July 20, 2008

I think I'm busted. I heard a racket on the south side of the house during the night, and I now have a pile of eight sticks lying right under one of the dummy cameras. The whacking on the side of the house was probably a Foot systematically lobbing sticks at the camera trying to activate it. Most likely they have now either figured out it doesn't work or they're wondering how it works without flashing. Dang! These rascals are smart! I figured with the dummy cameras AND chemically treating the yard so they can't approach the cameras would solve the problem for at least a week or so. It took them EXACTLY TWO DAYS. Guess it's back to the drawing board. I know who the dummy is now!

The Helena Independent
Helena, Montana Territory
October 28, 1882

A Wild Man in Idaho

Two cowboys who just came in from Camas prairie relate an experience which will probably go a great way toward re-establishing the popular faith in the wild man's tradition. On the first day of this month two cowboys, searching for cattle lost in the storm, passed over some lava crags and were startled by suddenly seeing before them the form so often described to them. They were so terrified that they sat upon their horses looking at it in dread. Mustering courage and drawing their revolvers, they dismounted and gave chase, but the strange being skipped from crag to crag as nimbly as a mountain goat. After an hour's pursuit both young men were so completely worn out that they both laid down, seeing which, the wild man approached them and stopped on the opposite side of the gorge in the lava, from which point he regarded the cowboys intently. The latter would not shoot, as they considered it would be unjustifiable, though they kept their pistols ready for use, while carefully returning the compliment thoroughly inspecting the phantom of Snake River.

The wild man was considerably over six feet in

height, with great muscular arms which reached to his knees. The muscles stood out in great knots and his chest was as broad as that of a bear. All parts of his body were covered by long, black hair, while from his head the hair flowed over his shoulders in coarse, tangled rolls and mixed with a heavy beard. His face was dark and swarthy and his eyes shone brightly, and he acted very much as a wild animal which is unaccustomed to seeing a man. The boys made all kinds of noises, at the sound of which he twisted his head from side to side and moaned—apparently he could not give them any "back talk" so, wearying of eyeing him the two boys fired their revolvers, whereupon the wild man turned a double somersault and jumped fifteen feet to a low bench and disappeared, growling terribly as he went.

It is supposed that this is the same Apparition so often seen before. The man, no doubt, does as the Indians did for subsistence and lives on Camas roots, and he no doubt kills young stock, as many yearlings and calves "disappear" mysteriously and nothing but skeletons of them are never found.

CHAPTER THREE

"When I Find A Special Spot . . . ": A Local Guide Appears, then Disappears

August 2007

I can just make out Sylvester's dim shape in the mist as he maneuvers his kayak. It's 6:30 in the morning and, in a borrowed canoe, I am awkwardly paddling toward a man I have only met once, Ojibwa by descent, who has agreed to take me to a forest he's been carefully exploring, and calibrating, for greater than a decade.

I've been rowing for twenty minutes and we're only halfway to our destination, along a winding beaver canal. The canal divides a wide swamp and is lined by reeds so high that, riding low in the water, one cannot see over them. Given the layout of the topography, and the natural buffer created by the swamp, navigating this canal is the most direct route.

I reach my new friend now and, after exchanging a few words, we paddle together toward the peninsula. There is no sound except songbirds, our breathing, and the gentle swish of water. "When I find a special spot," Sylvester has told me,

> what I always do is to put something somewhere so an observant person or whatever, would see it and wonder why or how it got there. I put two quarters on

> the ground about a foot apart. Right out in plain sight. My idea is that no one could resist a couple quarters. If they disappeared then I know I have people in my woods and should keep an eye and ear open for them. I also had them facing up and pointed them to face north. Nothing happened the first few years . . . till last year. Then a quarter was gone. I examined the area and found in place of the quarter a cute little arrangement of feathers. Each a different color. And spread out just like picture, facing east. If it were a person they would have taken both quarters and would not have left such a beautiful little arrangement, I imagine!

I feel, of course, extremely privileged to be taken in trust by this man. After having logged more than nineteen thousand miles on expeditions with the BFRO, I find it nearly impossible to believe, yet also delightful to contemplate, that my best bet may have suddenly announced itself not twenty miles from my home.

I first became aware of Sylvester a couple of weeks ago, when he submitted a report to the BFRO Web site.

> I hang out in the deep woods of Vermont a lot. I was out in the woods the other day and was walking thru the area when in the distance I saw a bright patch of ground. I couldn't tell what it was, but as I got closer I saw that it looked like a pile of hair. As I got closer to the flattened pile of hair I noticed that it was covering a layer of . . . what looked like huge poops. I looked around the area for more evidence, like a carcass, but couldn't find anything. I then very gently sifted thru the pile of hair for evidence of whatever it was with a small stick. I didn't disturb the area at all. It was indeed a layer of hair and under it was a layer of poops. The hair was interesting. It looked to me

> as if someone had shaved the animal. As the hairs were all cut even. So . . . a bear . . . with a razor or shears? A weird hunter that poops all over the ground then kills a deer and covers it with the hair?

I immediately contacted him, and that's when he let me know of his extended history playing sylvan games with "someone." Until recently, though, he'd never suspected that this someone could be Sasquatch.

He invited me to visit his house, one of several by the road in a small nearby village. Touring his upstairs music studio, admiring his collection, which hangs on the wall, of more than a hundred vintage swords, and then watching high-definition footage, on a ten-foot-diameter video-projection screen, of his forays into his "special spots," including his discovery of the broad mat of thick, white-gray hair or fur on the ground, I marveled at how many thousands of times I've traveled past this nondescript home, over thirty-five years, without the faintest inkling what it might contain.

True to his initial report, the video did seem to show clumps that had been *cut* from whatever body, and indeed, nearby, the sharp broken bottom of a green bottle. Sylvester's current theory is that with the arrival of warmer temperatures this spring, the Sasquatch, if this is what it was, sliced off its own winter coat using this makeshift "razor." After the lights came back up, he showed me a few of the tufts he'd gathered, and they seemed to match what he'd encountered on-site.

The Peninsula

In our two boats, we finally arrive at the muddy bank where he always runs his kayak aground, at the foot of an imposing, hundred-and-twenty-foot sheer face of pine, rock and roots. As he helps me to secure my canoe, he tells me, "We're in *their* woods now."

After a difficult climb, we crest the ridge and I see that its top is very narrow, twelve to fifteen feet across, and is covered by tamped-down pine needles: a clear pathway, with none of the punctures made my deer hooves. Sylvester says with a laugh, "I keep picturing them running along this ridge all night."

After we explore awhile, an odd general feature becomes apparent. This peninsula is probably three hundred yards long and a hundred and fifty wide, and the ridge top wraps around it to create what might be better termed a *rim.* Within this rim, the peninsula dips into a bowl shape to form a hidden valley. I see zero sign of human presence—no wrappers, tin cans, cigarette butts, water bottles, bottle caps, fire rings, etc.—and Sylvester says that in fact he's *never* found any such up here. This place is just too difficult to get to. The one exception, in fifteen years, is that broken bottle, tossed near the white-gray hair. And even this, he found on his next visit, was duly removed from the premises; most of the hair, too, had been cleared away.

I say "peninsula," because this is its shape, even though the formation rises not from open water but out of swampland.

As we sit for a respite on the rim, he shares some personal information. Native to Wisconsin, Sylvester was adopted as an infant by an Anglo couple and raised in England. He ran away from a dreadful home situation when he was sixteen and married at eighteen. (He's been with Ellen now for thirty-two years.) He never met his birth parents, and they are both deceased. His shoulder-length, graying hair and broad, earnest face drip with sweat as we recommence our tour of the peninsula, clambering over fallen trees and up and down steep slopes made nearly impassible by logs and limbs strewn everywhere, and by thick undergrowth. "If it's really tough to get to," he says, "that's where we should go."

He tells me that by laying low in certain clefts within this peninsula, he's been lucky enough to hear bipedal footfalls, other sounds he can only describe as "jaw-popping," and what seems like women talking, with distinct syllables but unintelligible words. Once, he says, he was easing along the canal in his kayak and, just as he rounded a bend, dipping his paddle into the water with a slight splash, he heard something huge powering across the waterway and storming up the bank to his left. Then, it sat up there, out of sight within foliage, "grunting and grumbling."

This secret peninsula is less than three miles from where, in 1991, my good friend Tim, an extremely level-headed man, a writer and college administrator, witnessed something bizarre. Late one night, he was driving with a female passenger when they rounded a bend and saw, caught in the headlights, a small figure standing upright. "It was maybe three, three-and-a-half feet tall, no more than that," Tim told me. His passenger said, "Is that a . . . *poodle*?" This is the only association she could make to this vision, a dog standing on its hind legs. Tim saw, instead, a monkey. "It had a little leathery face, and it was just staring at me, looking at me sort of over its shoulder, frozen in mid-stride. And then, all of a sudden, it ran off the road and into the underbrush and trees."

Stick Parties

Sylvester points out several fascinating manifestations, some new even to him, altered since last he visited just six days ago. Some stick and tree structures have, it seems, been laid waste: "Wow, they've totally destroyed these! I have pictures from right here and everything looks different."

And there are shallow trenches covered by little apparent "roofs" woven of sticks and leaves; these he has found before, but now they have been fortified by large leafy limbs set about like barricades. Maybe, he speculates, he got too close, should not have touched the roofs?

But by far the most intriguing and useful lesson Sylvester teaches me on this first foray, and indeed on our few subsequent trips into the woods, before he decides to pull the plug on our association, is what he calls "stick parties." These are spots where branches have been snapped into short segments, by the dozens, six, eight, ten inches in length, and in many cases the bark has been stripped as well. Many are the jumbles, like the old game of "Pickup Stix." He trains me to notice these, saying, "Yeah, *that* sure looks naturally occurring."

Once you start spotting these, of course, it's hard to stop, and you can come to doubt your judgment, but it's possible to learn to rule out random deadfall/windfall. The sheer number tells the story, along with the bark-peeling and the fact that the sticks have been broken into short segments. Also, they will often be improbably wedged into tight spaces between rocks or trees.

Sylvester writes me all excited, two days later, having just experienced a breakthrough. "I used my hearing device [a microphone augmented by a parabolic dish] and what a revelation. I listened to one eating sticks. It was fascinating. I could hear it eating and gathering more sticks. My sense was that it was standing sentry, keeping tabs on me. The sound came from two hundred feet away, or so, and the source of the sound stayed out of sight."

Researching on-line, I find this: "Why do animals eat the bark of trees and fences? They are looking for minerals. Tree roots go deep into the soil to absorb minerals, which are then present in the bark." And then I locate an article (by Ker Than) about wood-eating behavior in Ugandan gorillas.

> After observing mountain gorillas in Uganda for nearly a year, scientists believe they have discovered why the animals eat wood and lick tree stumps, behaviors that have puzzled primate researchers for decades. The answer: for the sodium. Gorillas in the Bwindi Impenetrable National Park in Uganda will suck on wood chips for several minutes before spitting them out. Sometimes

> they chew on them until their gums bleed. They have also been seen licking the bases of tree stumps and the insides of decayed logs, and breaking off pieces of wood to munch on later. Gorillas will return daily to the same stump and take turns feeding.
>
> A new study by Cornell University researchers potentially solves the mystery. The researchers observed fifteen gorillas of different ages and gender as they engaged in wood-eating activities. After the animals were gone, the researchers collected wood samples from stumps and logs that the animals consumed as well as those they avoided. They also collected samples of other things the gorillas ate.
>
> The researchers analyzed these items for their sodium content and found that the decayed wood was the source of over ninety-five percent of the animals' dietary sodium, even though it represented only about four percent of their food intake.

I eagerly share this new connection with Sylvester, yet he's mildly interested at best; I'm coming to learn that he's a fiercely independent researcher, cares about ideas only if they're generated by himself, and I get the sense that he's quickly losing interest in my contribution.

Crime Scene

Ten days later, we meet up in another forest, several miles from the peninsula. I'm happy because we can drive here, rather than paddle, and we proceed down a very narrow dirt road, just after dawn. Although he won't put me in direct touch with the witness, Sylvester tells me that he's recently spoken to a local man who says he saw a tall hairy figure crossing this road late at night, and that everyone around here avoids this area, believes it haunted. There are screams in the night.

I recognize this road from my youth; as a teenager, I used to go running here. "I think a logging road shoots off to the left," I offer, "just ahead." Indeed, there it is. We pull over, tuck the car out of sight at the trailhead, cross a field, and enter the woods.

We're immediately transfixed. It's like a classic Sasquatch cathedral or playground in here. We see stick parties everywhere, dead ringers for those present on the peninsula, peeled and strewn, as well as peculiar, enticing structures: barkless branches and limbs leaned carefully up against tree trunks; whole trees pushed over into diagonals, here and there, alongside vertical trees, the *same* diagonals, gentle, like this /; and high concentrations of sticks all shoved together to form mounds.

Then we find fresh deer bones littering the ground, leg bones, linked vertebrae, an eighteen-inch skull, all representing, it seems, several different kill sites. Near the skull, we are stunned to stumble upon a great heap of *hair*—coarse and curly, brown, like an absurdly heavy-gauge wig dumped on the ground. Sylvester gathers it up, smells it, and is overwhelmed. "It's so musky, so . . . *animal.*" I smell it too, and he's right. Some of the curls pull out to fourteen inches in length.

We're both feeling graced and giddy by this discovery, and we pack it off home. Actually, *he* pockets it, allowing me just a very limited hank. Later, when I hold my share up to the sunlight, an iridescent red—like henna, like roan—clearly emerges, and I'm reminded of all the hundreds of witness reports that describe Sasquatch as reddish-brown.

That Sylvester has kept 98% of the hair for himself does sort of irk me. But then again, he's taken on the role of leader in our relationship, of shaman guide, and I've tried to go with the flow. Certainly, I have learned much from him.

The next morning, I return to the site to take pictures and video. He was the cameraman yesterday, but hardly took any shots, so our visit was underdocumented. Also, I'm trying to cast us, instead, as equal partners, and so I immediately report my photography trip, email him the pictures. Big mistake.

His reply is too angry to reprint here, and then it takes days for me to extract a more composed critique.

> What I do Chris, is "magic". I am directed to what I want. Then it is up to me to figure out the next step. When? When I am directed to do so. Most of the investigating isn't about pics and video (although it doesn't hurt). We are RESEARCHING this thing. It is not an archeology project. We are dealing with forces etc. that we don't understand yet!
>
> I go into an area to gather intelligence. Thru sight, sound, general feel. Then the whole picture comes to me, gradually. When i am 99.9% convinced, then i go from there. If I need to return to the area I know that the same vibes will be there again, plus the new ones that are given to me, from that trip. Psychic research is the key to all of this stuff. I can't rely on just my instincts. I am dealing with something that doesn't play with the same rules we are taught. So I use my gift to unravel truths.
>
> I hope you can understand what I am saying. This is not finding ruins! This is like a "CRIME SCENE" Investigation. There is yellow tape all over where we have walked. This means that this is not to be disturbed unless it is done methodically. What we do is gather intelligence on each place and then expand our treks deeper into the areas. If we decide to go back into an area we don't disturb the SCENE!

> I really do appreciate your enthusiasm. I like how your mind thinks. (when you're using it).

Needless to say, it is not long before he drops me altogether, like a stone, but now I'm armed with what I consider powerful new knowledge, thanks to his fleeting mentorship. After all, without it, I'd probably never have found out that my own backyard is a "hot zone."

I send Jerry several strands of the hair Sylvester and I found, and he says yes, it looks just like that of his new acquaintances in Upstate New York, and of the friends of his youth downstate—brownish-red and frizzy.

Taking up, then, my solo local research, I studying satellite maps and realize that there is, adjacent to the area with the deer bones and the hair, a steep ravine that bottoms out into swampland. Suddenly, it dawns on me that I *know* this ravine! When I was fifteen, I found it not far from my home, and explored it often. It always exerted a kind of sacred gravity for me, seeming a realm *apart*, and I'd return from time to time, in adulthood, to replenish myself on its singular atmosphere.

Today, I hasten back, and while it strikes me in the same primordial manner, I see it now through a whole new lens. Over the next two months till snowfall, I spend more than fifty hours inside the ravine, talking and singing to the trees, making wood knocks, and just sitting quietly. I can hardly believe what it is granted me to experience, kicked off by a crashing tree on a still night, and a roving tomato.

Woodland Daily Democrat

Yolo Country, California

April 9, 1891

An Unheard of Monstrosity Seen in the Woods Above Ramsey

Mr. Smith, a well-known citizen of Northern Capay Valley, called on us to-day and tells us the following strange story which he would be loath to believe were it not for the fact that he is an old acquaintance of this office, and has always borne a spotless reputation. Several days ago, Mr. Smith together with a party of hunters, were above Ramsey hunting. One morning Mr. Smith started out early in quest of game, but he had not gone far when his attention was attracted by a peculiar noise that seemed to come from up in an oak tree that stood near by. Looking up Mr. Smith was startled to see gazing at him what was apparently a man clothed in a suit of shaggy fur. Having heard of wild men, he naturally placed upon his guard, but thinking that he would see "what virtue there was in kindness," he called to the supposed man to come down. The speech did not have the desired effect, rather the opposite, for the strange thing gave grunts of unmistakable anger. Our informant went at once in a bee-line for the camp. After placing some distance between himself and the strange creature, the hunter turned around just in time to see it descend the tree. Upon reaching the ground, instead of standing upright as a man would, it commenced to trot along the ground as a dog or any other animal would do.

Smith then realized that it was no hermit he had seen but some kind of monstrosity. The hunter stood amazed and spellbound for a moment, but soon gathered his scattered senses again and was soon making his best speed to camp, where in a few breathless words, was telling his companions of what he had seen.

They were disposed to laugh at him at first, but his sincereness of manner and his blanched cheeks soon proved to them that he had seen something out of the usual order of things.

A hasty council was held, and the party decided to go in search of the monster, so taking their guns and dogs they were piloted by Mr. Smith and they soon came in sight of the unnamed animal. In the meantime it had commenced to devour the contents of Mr. Smith's game bag that he had dropped in his hasty retreat. The creature would plunge its long arms into the bag and pulling forth the small game, transferred it to its mouth in a most disgusting manner. An effort was made to set the dogs upon it, but they crouched at their masters heels and gave vent to the most piteous whines. This attracted the attention of the creature, and it commenced to make the most unearthly yells and screams, at the same time fleeing to the undergrowth. The whole party immediately gave chase and soon gained upon the strange beast, and it, seeing that such was the case, suddenly turned, and sitting upon its haunches, commenced to beat its breast with its hairy fists. It would break off the great branches of the trees that were around it, and snap them as easily as if they had been so many toothpicks. Once it pulled up a sapling five inches through at the base, and snapping it in twain, brandished the lower part over its head. The hunters seeing that they had a creature

with the strength of a gorilla to contend with, beat a hasty retreat.

Mr. Smith describes the animal as being about six feet high when standing, which it did not do perfectly but bent over, after the manner of a bear. Its head was very much like that of a Human being. The trapezie muscles were very thick and aided much in giving the animal its brutal look. The brow was low and contracted, while the eyes were deep set, giving it a wicked look. It was covered with long shaggy hair, except the head, where the hair was black and curly.

Mr. Smith says that of late sheep and hogs to a considerable extent have disappeared in his vicinity and their disappearances can be traced to the hiding place of the "What Is It."

CHAPTER FOUR

Applying the Lessons in My Childhood Forest: The Vermont Ravine Project 2007 & 2008

One day in 1973, at the age of twelve, while running on a dirt road along a ridge above my house, I decided to veer off onto an old, grown-over logging road. It took me through a corridor of pine trees, then around a sharp corner. I felt like I was entering Narnia. The road formed a downward-slanting shelf at the edge of a deep ravine. I ran for a minute or more, tracking downward, till arriving at a marshy bottom.

Immediately, I was entranced by this space that seemed sequestered from the ordinary world. I frequently explored the near bank, admiring the far side, a steep wall of thick trees, but never crossing over there and climbing it. This ravine stuck with me indelibly so that, later on, I adopted it as a sacred space in my first published novel, in 1988. There, I called it "the Scoop-Out" and remodeled it into a field where my young couple would meet to compare notes on life, escaping utterly from everyday concerns.

> Baker hadn't expected anything to happen. He'd just wanted to show her this place that seemed to him

different from any other, where no one else had ever been. She was quiet as they turned onto the little path he called Time Lane. This path always seemed to take him back to before people knew how to talk. (*He'd* never talked in the Scoop-Out. Anything he'd ever thought to say seemed stupid when there was this high-strong insect sound forever playing its one note down in the field, bright to hear.)

Now that he was watching another person run in this place (she sprinted up the banks . . .), he could get a picture of the whole Scoop-Out as a big soft bowl. The field was hollowed out of hills and ridges on all sides, and on the banks was moss, thicker and smoother, without any bare patches, than any other moss he had ever known.

Late Summer & Fall 2007

Less than a mile, as the crow flies, from where I spent my adolescence—occasionally daydreaming of the creature I'd seen in The Patterson Film, wishing I were grown up so I could travel to the Pacific Northwest—I begin, finally, at age forty-six, a long series of forays into the ravine. Luckily, it's still undeveloped, rugged and pristine.

On my first visit, I stick to the near side, near the pubic dirt road and just a quarter mile from a (human) residence. The same old logging road proceeds downward, and here I notice that two trees have been pushed slantwise across, so that I need to duck under them. I also find, impressed into the dry mud, several large impressions, potential tracks, which I duly videotape.

At the bottom, several hundred feet down, I look across the swamp to the far side of the ravine, the unpopulated side, difficult to access. What's awaiting me over there?

On my second visit to the near side, three days later, something new: two birch saplings, maybe two inches in diameter and twenty feet long, have been sculpted into arches, bowed across the trail and pointed down, such that their leafy tops dangle in the precise middle of the path. This is astonishing, a definite sign.

Upon examination, one of the trees proves even more interesting, because the dirt beneath is swept clean. I mean *really* clean, starkly different from the rougher texture of the earth surrounding. The least radical interpretation of this is, of course, that the wind has simply stirred the leaves around, which brushed the turf. But down here inside the ravine, there just isn't enough wind to agitate such a tree so effectively. (And why was it bowed over in the first place, between one visit and the next? No snow and ice storms in September, even in Vermont.) And as I imitate the wind with my hand, I realized that it would actually lift the leaves *away* from the ground. More likely, I reason—though, yes, the word "likely" still seems ludicrous to me in this context—the Sasquatch who bent this sapling all the way into a semi-circle found it necessary, in order to make the new trajectory stick, to continuously yank it down into the ground, to really *thrash* it about. Was it more stubborn than the other tree?

At the bottom, I walk farther along the trail than the last time, and here I find that two more trees, six or seven inches in diameter, much thicker than the saplings above, have been snapped cleanly in twin right angles and laid, a clear No-Trespassing sign, from right to left across my way.

Trespassing, I find a strait across the swamp to the far side of the ravine. This bank is far steeper than the other, and a hard climb takes me eventually to several level tiers. The forest over here shows no sign of human trails or refuse. Soon enough, I come across a very obvious and exciting T-pee stick structure, as well as many other little stick party scenes of the exact sort that Sylvester inculcated me to.

Also, I see many more bowed young trees, some with their ends tucked and anchored carefully beneath older, fallen trees. At one spot, *two* are arched and pinned right next to each other, one beneath a long fallen tree, the other beneath *another* fallen tree, shorter, that has been, in turn, pinned beneath the *first* fallen tree.

But even more remarkable, the T-pee structure employs two bowed saplings as well, both still rooted in the ground. One enters the formation, pulled through from ten feet away to become part of the weave at the top of the T-pee, where the slanting sticks converge, and the other, growing within the T-pee's radius, passes up through the structure, and exits, bending away through the air. Amazing.

I find my own "special spot" beneath a pine on one of the terraces between the steep lower portion of the ravine and the steep upper portion. One hundred and fifty feet above me, I can see the clean line of ridge through deciduous trees, and I decide never to mount this wall, but leave it for *them*, a strategic overlook, a safe vantage.

The Third Tomato

To emulate Jerry of New York State, after unfolding my chair, I set about constructing a simple lean-to out of smaller pines I find lying on the ground. As with many small- and medium-sized pines in the area, these appear to have been felled by force—their trunks end in a ragged mess, not a smooth sawn surface—and then dragged from their sources. I count seven such instances in the immediate vicinity, and nowhere do I see the stumps.

My first order of business, after making my structure, to inaugurate my site, is to place an offering of three tomatoes from my garden. I wedge them between two close trunks, where they glow nicely in the sunlight.

Apparently, not only do Sasquatch not appreciate tomatoes, but neither do other creatures. When I return the next day, nothing is changed; and when I return again, a few days hence, one of the tomatoes remains in place, one has softened and fallen to the ground. The third I cannot locate at all, but I assume it must have fallen as well, and rolled somewhere among the old leaves. I figure I'll look for it after sitting and reading for a while.

After about an hour, I go to pee at a tree I've used for this purpose several times before. Exactly at eye level, I turn to see none other than the missing tomato itself, just inches from my face. It's been slung over a branch, sitting in the crook. I say "slung" because it's much farther along in the decay process than the other two, and whereas their skins are still smooth, this one's is pocked and mottled, seeds visible, the whole thing now pulpy enough to wrap partway down around the branch, a certain liquefaction having already set in. Has it perhaps been inside a mouth before being rejected and left here for me in "the bathroom"? And that it's not been bitten into pieces, even two pieces, would say something about the *size* of said mouth.

It feels like I have suddenly, in effect, lost my virginity as a researcher, because this tomato seems meaningful, and *meant*. I've now joined the gifting and receiving ranks of Sylvester and of the other habituators.

Yet none of what I leave at my site, next, elicits any further such reciprocal gestures. I wedge apples between the two trunks, and they are always taken. I suspend an apple from a limb nine feet off the ground, secured by a tight noose of picture-hanging wire, and it too is gone when I return.

Between August 31 and November 2, I visit the ravine twenty-nine times, staying for an average of two hours, talking to the woods, reading, making my plaintive wood knocks and whoops. After each trip in, I exit the forest by the same route, passing finally through what I call "the foyer," a margin of needle-less, light-deprived pines that stands between the ravine proper and the dirt road where I park.

Further Signs

One day, on my way out, I'm greeted by a mushroom. Small and white, pristine, tilted, and again just at eye-level, it is propped within a fork. There are no mushrooms growing on this tree, and a search of the ground at the base shows no little grove of same. It's been plucked and carried from somewhere else. I lean toward Sasquatch, but for some reason I don't take this gesture quite personally, don't *accept* it but leave it alone; that is, I think of it as an interesting case of dainty "accessorizing."

A few days later, and maybe a hundred feet further inside the foyer toward the road, beyond the mushroom (which has disappeared), I'm met by a treed apple, suspended at neck level inside a cluster of pencil-thin branches. I recognize it as likely one of those that I left back at my site; there are no orchards in the area and it has the same pale, varicolored skin. A bite has been taken.

This one I do take personally—I'm thrilled!—yet because I'm still a novice, again I don't think to join the game, not here. Back where I sit, I've been primed to play along, and robustly. All it takes, apparently, is this subtle shift in the expected terms, this sleight-of-hand, to throw me off.

I regularly explore the vicinity, finding many more suggestive stick structures all within a half-mile radius. On one occasion I glance down a forested hillside and do a classic, comic double-take. I squint. Probably not meant for me, but a grand gift nonetheless—a striking, artful, gorgeous "pinwheel" formation. I slip and slide down toward it, heart accelerating, then approach like one of those monkeys at "the slab" in *2001: A Space Odyssey.*

It's constructed of four long, straight trees, crossing at the "hub," two of which are still rooted and have been pulled through in a manner similar to the arched saplings but also different, in that by the time they enter the "wheel" they are not much bent anymore but act as proper "spokes." In fact, it takes time for me even to recognize that they are still growing and have been, in effect, borrowed for the cause.

In the picture I take, one can see a typical thin arched sapling in the foreground. When I return a week later and take another picture from the same angle, this tree has been snapped at the top. I do not know how to understand this, except that perhaps I'm being told my discovery of the pinwheel has not gone unnoticed.

First Night in the Ravine

Not brave enough, yet, to spend the night here on the far side of the ravine, I try the near side, closer to the road and houses, closer to civilization, closer to my get-away car. I use no tent, because I don't want to feel cut off. Brave, this? Not really: I figure it will be tougher to sneak up on me if there's no thin vinyl wall between "it" and my ear. Not only that, but I choose a spot three-quarters of the way up the logging road, so that I might escape to my car, if need be, hurtling through the blackness.

I arrive at 10:01 and set up my high-quality, newly purchased audio recorder, hanging the microphone over a tree limb maybe six feet from my head.

I settle in, getting warm, telling myself that this whole *fear* thing is a bit ridiculous, after all, that I'm just here on a simple listening mission, to document the action, if any, occurring a quarter mile away, over on the far side, where they seem to live.

It's an utterly still night, and forecast to fall to the mid-forties: perfect acoustics. I've almost successfully persuaded myself into a comfort zone when it happens. From up on the hill behind me, the way I came in, between me and my car, comes the abrupt CRASH! of a tree brought down.

The mind is funny. I am definitely freaked out but I instantly begin the project of believing that the tree could—couldn't it?—have fallen of its own accord; maybe it was just ready.

Only at dawn, when I wake and pack up, do I begin to reckon the long odds of any tree falling, with such loud force, when there's no wind; that is, if the tree was that weak already, wouldn't it have been toppled by the previous storm?

But even more to the point, a hundred feet up toward the car, I am confronted by the fact of precisely *where* the tree came down: partway across the trail. Of all the hundreds within earshot, this is the one that decides to crash down onto a path with a history of barriers, the two leaning trees here near the top; the two bowed saplings that appeared for my second visist; and, at the bottom, those two snapped into approximate right angles, barring the way. Also, this location would afford the perpetrator a quiet, grassy approach, avoiding the crunch of dead leaves, so that he could take just one step up onto the shoulder of the logging road, reach out, and yank that tree down.

Later, I listen carefully to every second of the hours of audio recording. Most of the time, I am contentedly snoring, and hearing this—especially after having wised up to the probable nature of the tree-crash—feels eerie, vaguely sickening, actually, forced to encounter oneself so exposed in the night. And then there are the other sounds. At 2:14 AM, an odd thump, a series of very long howls coming, it seems, from the far side; at 2:31, something ticks the aluminum of my parabolic listening dish, which I'm learning to use to augment the microphone. Throughout, possible footsteps, but if footsteps, extremely careful, like tip-toeing. Also, occasional twig snaps. Evidently, something is sneaking around.

Listening in the safety of my home, I remember the dozens of accounts I've read and heard, of people camping out and experiencing initial threatening demonstrations, such as limb breaks, whole trees being shoved over violently, even bipedal bluff charges, all clearly intended to freak the humans out and chase the intruders from the territory. Typically, this works. But if these people are, say, Sasquatch researchers (alone or on group expeditions) and not willing to vacate, what often occurs is that much later at night, after 2:00 or 3:00, when they are most incapacitated by slumber, curiosity takes over and the Sasquatch will sneak near. Audio recorders will frequently pick up approaching and retreating footsteps in this time frame, or else the people will actually awake to the sound of a hand touching their tent walls, and sometimes, if there is a strong moon, they may see a massive silhouette.

Exactly one week later, I return to the downed tree to find that another, a birch, has been laid across the first, at an exactly perpendicular angle, to form an X.

Yet Another Tomato

One morning in early October, a house guest brings me something noteworthy. At the edge of the pathway leading to an A-frame cabin at the back of my property, he found a tomato. Likely, it came from my garden, out front, but how could it have traveled three hundred feet? I'd have guessed raccoon or squirrel, had the skin been bitten or punctured, the object's shape altered from the round. We examine it carefully: the skin is all unmarred.

And even so, I'd probably still have passed off this misplaced produce as some unaccountable quirk of nature, if "tomato" had not recently become a meaningful currency of exchange in this game I seem to be caught up in. I pay attention; after all, my property is only just over two miles from the ravine.

And I have not mentioned that incident to this man. The night before, he argued against the possibility of Bigfoot ("I just can't *see* there being a gigantic primate being able to elude us like that, y'know?"), but this morning, he's ready to pronounce his discovery "definitely odd."

Man oh man, I like this hobby!

Of course, as with so much that occurs in the realm of Sasquatch study, one does find oneself suspecting an insidious slide into mental illness.

One proceeds.

Spring 2008: The Adjustment

Between late April and early November, I spend twenty-five nights in the ravine. The majority of these bring nothing but mosquito bites, birdsong, and lonesome darkness. Or nothing obvious, that is, because of course one can never know when habituation might be secretly underway—in both directions, including my accustoming myself to their nocturnal world.

The first six nights I sleep at the base of the ravine, beside the swamp. The idea is to give the Sasquatch maximum strategic advantage over me. I'm remembering Dian Fossey's observation, "On a slope gorillas always feel more secure when positioned above humans," as well as many accounts of Sasquatch encounters in which the primate makes its way downhill toward the overmatched humans.

By mistake, I use a tent, muffling my relationship with the surrounding woods, and of course I sleep soundly, uninterrupted. Nor does the digital recorder (hung in a nearby tree) pick up anything interesting. Once, a porcupine happens through camp and, starved for company, I lurch outside and shine my light into his little, nonplussed face.

Each evening, at dusk and on into serious dark, I play a few minutes' audio of my toddler's voice, her just babbling, laughing, being silly, testing her range. I'm trying to follow the suggestion offered by Robert W. Morgan (see Introduction) to "create a provocative routine . . . .

> Find a way to gently announce your presence. Try whistling or singing a tune now and then. Why are you doing this? Because that music will serve as your trademark with them, and when they hear you play it, they'll know who you are.

I point the CD player upward, balanced on a branch, liking to imagine my daughter's reedy, enthusiastic tones mounting the bank, all the way up, fourteen hundred feet through gathering night, to strike an ear at the top, primordially geared to respond to the very young within the clan.

Back home, even before the snow melted, she and I began to leave fruit, pretty quartz stones, and, like the Oklahoma family, adorable stuffed animals, at a gifting station down behind our house, a spot chosen because that's where the garden tomato was mysteriously placed last fall. Also, there are two dramatic tree arches several yards away, each one anchored emphatically. Once, in late April, an apparent wolf, remarkably nearby, howled at us, loud, long, and lovely, while we restocked the gift basket (apples having been removed). I was struck that this creature would approach us, and sound off so boldly, and in broad daylight.

On June 9, I decide to abandon the tent, this whole low-altitude project, and scale instead to my original daytime post, from last year, by the switched tomato. It's a mile hike from my car, part of it so steep that my heart quickly reaches that stopwatch tempo. Why it took me this long to refigure my position I'll never know, but the change pays immediate dividends.

Keyed up and, sure, okay, frightened, requiring some assistance, I pop two Rite-Aid generic sleeping tablets. I can excuse this drug-use as scientifically mandated, given that, historically, snoring seems to put them more at ease.

I wake in the middle of the night, sometime after 2:00 AM, quite confused and facing an impenetrable wall of darkness. I'm slapping mosquitoes and hearing distinct stick breaks, a hard-to-describe picking or rummaging through undergrowth, plus occasional thumps on the earth, somewhere generally "over there," spaced by long intervals of silence. As during my overnight on the other side of the ravine, eight months ago, these thumps are isolated, never doubling up into any clear, bipedal signature.

Until 2:35, I'm able to pass these sounds off as possibly just my desirous imagination projecting itself onto a neutral background of pinecone falls and small-fry rodents. Maybe a deer snapped the twigs with its hoof.

Suddenly, though: a loud and intricate birdcall from the exact direction of the stick breaks. This is hard to reconcile with the image of a deer. I remember the habituators' reports of such unlikely, dead-of-night songbirds.

I sit up and stare into this blind nothingness, feeling a peculiar mix. On the one hand, I'd expect myself perhaps to freak out, since the way to the safety of my card is rugged, the downgrade inviting missteps and disaster, and yet, on the other hand, my worst enemy is, after all, *actual* nothingness, being stranded and truly alone up here (or at least the only primate), on some fool's errand; in this sense, *any* sounds, especially such promising sounds, are a welcome affirmation. Many have asked me, "Aren't you terrified out there by yourself?" and I've answered, "Well, my fascination always outweighs the fear." This rather glib response has the advantage of being, for the most part, pretty accurate, and never more so than when, this first long night on the high terrace, I'm privileged to be woken up by a nearby double wood knock, maybe sixty feet away, maybe eighty, hollow, concerted and clear: *thwock thwock*! I smile under my blankets, glad to be tentless and feeling included in some far practice, foreign yet intimate. I'm strangely at home, nowhere in the world I'd rather be. 3:29 AM.

The subtleties continue night by night. June 15 brings a dramatic visit, apparently by the same wolf (or a dead ringer for that wolf, or a Sasquatch imitating a wolf) that sang to my daughter and me back home. Just at dusk, a gruff series of barks leading to drawn-out howls, this sequence repeating again and again for three minutes and twenty-nine seconds, the howls sometimes trailing off into a definite "woooo-oooo." Oddly, at intervals, an owl (or an "owl") responds throughout the howling, from maybe twenty degrees removed on the compass circle.

Whatever the source, this serenade scares but honors me. I am concerned that if it's really a wolf, and if I maintain my method of surrendering myself by means of Rite Aid sleeping tablets, it may be able to tear open my throat before I can wake and defend myself.

For a few days, I contemplate using a cervical collar to sleep in but end up wrapping my neck in a folded towel.

Thumper Night

I fall asleep early, June 24, partly because there's nothing to look at. It's extremely dark thanks to a thick leafy canopy and, beyond, cloud cover and no moon.

At 9:51, I become aware that I'm not alone, and that what's with me must have hands. What I hear is a snap-crackle-pop of sticks being deliberately broken, emphatically and close by, maybe fifty feet away. It feels exactly like the Arrival I've hoped for so fervently, but I never expected I'd have zero eyesight like this. I possess no night-vision technology, nothing but a large hand-held spotlight, and using the latter is of course a well-known taboo among researchers. The idea is to build up trust.

The sounds go on and on. Sometimes three sticks will be snapped within five seconds. The more I listen, the clearer it becomes that this display is *moving*, slowly circling me. Best I can tell, it's just *one* of them. *Just*? Unless a person has hiked in here and found me in the pitch dark, when I've told nobody where I'm camping, and has decided to behave like a Sasquatch, this is actually a Sasquatch, and an old suspicion returns to strike me with fresh relevancy: We're told that though they may threaten us, they'll never hurt us, that it's not in their nature. Logically, this has always made sense to me, because it would not be in their interest to aggress against us; it would be anti-stealth, would "out" them, make them a target.

And yet, doesn't every species feature individual differences, a bell curve of traits and aberrations? In other words, in the case of any given confrontation, can't it be that *this guy* happens to be insane, has a brain tumor, a chemical imbalance, or is a teenager with severe emotional problems?

"Hello over there," I call out in my kindest voice. "Hello hello hello, I'm glad you're here. Please stay. What are you trying to tell me?"

From my backpack, I pull my daughter's one-piece pajamas, which she's outgrown. Invisible at the moment, they're white with blue polka-dots. I hug them to me, like some talisman against manual decapitation.

Hours pass, and the message remains the same, though I cannot read it. Occasionally, to the snaps are added light thumps. I keep talking, anything that comes into my head, whistling, even singing songs such as "Twinkle, Twinkle, Little Star" (my little girl's favorite). And then, shortly after midnight, these thumps become, for a short time, anything but light. My visitor suddenly smacks the ground four times, *pounds* it; half a minute later, three more.

I say, "I *hear* you over there." My tone is casual, jovial, and in retrospect, over the succeeding weeks, I'm increasingly surprised at myself, at this apparent lack of fear. I say, "I'm going to call you Thumper! Would you like an apple? I'm going to roll it over to you, okay?" But this elicits nothing further, certainly not the ideal response of the apple being rolled *back* to me.

In fact, never for the rest of this protracted night does he allow himself so much abandon again, but happily, the audio recorder captures the whole encounter distinctly.

By 1:35 AM, my composure is slipping. The stick breaks are impressive in their sheer constancy. I have not slept and my tone has gotten downright plaintive: "Well, don't just keep circling me. You're going to scare me. Why don't you just say something, or make a knock or something?" There's no way to discern the intent, here, but though initially I extended the benefit of the doubt, now I'm no longer quite so sure; I've started to feel menaced. I'm tempted, of course, to go to the spotlight, but I don't want to ruin this interaction, no matter what it is precisely trying to accomplish.

Between 2:00 and 3:30, I notice a tightness in my chest and jaw, but can't tell if I'm being mildly "zapped," or if instead these are the garden-variety anxiety symptoms one might expect when being surveilled all night by a legendary primate.

At 3:57, just pre-dawn, I stand up stiffly and lurch off into the night, my hiking boots landing on sticks that sound like gunshots going off. I'm holding the spotlight but waiting, thinking to flush the creature so that I'll know where to aim the beam. (Listening to the recording later, I *think* I can hear footsteps running away.) After twenty seconds, I jettison this tactic and switch on the light. Scanning, the bright circle looking paltry, revealing nothing but trunks and branches, I feel like a complete fool. I've just lost the game, and maybe the whole project.

After sleeping half the next day, I reach out to the wise women of the habituators' forum, and Ammi (in North Carolina) writes,

> Great night!!
>
> They must be getting used to you, to even make any noise to let you know they are there.
>
> They can circle yer butt all day, and not make a sound . . .
>
> You may need to be a bit more gentle, and not have the voice of someone who is working a deadline . . .
>
> They may not have our technology and tool use skills, but they have had just as many thousands of years, to simply hone the "read people" thing. Consider them serious specialists in that field . . .
>
> I am sure our concept of body language and voice tone, as a clue to real thoughts, is a joke to them. You have to understand, they don't just hide, they hide the fact that they exist, to most of the world.
>
> That means they don't make sounds, leave footprints, etc . . . . if they are doing this openly, you have to realize this was their first show of trust. Just to come out around you.
>
> I don't believe they are "stupid curious" . . . they seem to know just when, with whom, and how much, they can get by with.
>
> I think they test us like that.

Three Gifts

Over the following week, I worry that I have managed to destroy, through ego and lack of respect, all that I've been able to build—that *both* sides have invested—in this habituation enterprise.

And then, on three successive mornings, June 30, July 1, and July 2, I find waiting for me in the grass by the passenger side of my car, in the exact same spot each time, certain objects.

First, a large bulbous . . . what? It's shaped like an eggplant but it's partially decayed and lacks that rind. A fungus? Later, my neighbor recognizes it as in fact an eggplant; it came from her compost pile.

Second, a two-inch length of jawbone, the back portion, with all teeth in place. The teeth are not the flat grinders of a herbivore, such as a deer. A canine?

Third, a skull, apparently a deer's, with a little curl of greenery (still green) tucked into the brain cavity.

I have a dog and so does my neighbor, but what dog carries a rotting vegetable? Also, these two bony offerings are entirely desiccated, nothing whatsoever to gnaw on. This would make sense as a "gifting spot," furthermore, because a) my car is the means I take to travel between my home and theirs, and b) the passenger side faces away from my bedroom window, so that the things could have been left, by someone crawling, hidden, across the yard.

Okay, so if these *are* gifts, what's their meaning? Am I forgiven for losing my cool the other night? Or is the entire paradigm of the dire taboo perhaps a misconception? There is another angle from which to view this important night, rather than as a potentially game-breaking betrayal of a delicate trust relationship, and that is that what occurred is well *within* the rules, that the game is more in the vein of a good-natured sparring match, and that what I'm guilty of is no more than forfeiting one round among many. In this frame, the gifts would seem an affirmation, meaning: "Well played."

Maybe, even, I ought to think of turning on the light, at least a little each night, as less a brazen salvo than a methodological obligation, a ramping up in the habituation process.

And yet, I decide to err on the side of judiciousness, and to lay down my weapons.

Wood Knocks

The late evening of July 12 breaks out in light percussions, rather far away from down to my east, across the ravine, but definitely wood on wood, a long series between 10:10 and 11:00 PM. And though distant, they register clearly on the audio.

And then, the next morning, in that same direction, and on my route back through the ravine to my car, I notice another mushroom, balanced in the branches of a tree two hundred feet from my sleeping site, at eye level, just like the one I found last fall, even though back then I used an entirely different route, hadn't found the current hiding spot for my car.

The knocks of July 12, however affirming, are as nothing compared to those that pound at me two nights later, just at dusk. I've just relinquished both my hand-held spotlights, setting them almost ceremoniously at the base of the tree that held the new mushroom. "I won't be using these again," I tell the woods around me.

Twenty-one minutes later, as I'm settling in for the night . . . BANG! BANG! BANG! They come from back along last year's route, from where I heard the wolf or "wolf" nearly a month ago, that is, on *my* side of the ravine, and much nearer this time, less than two hundred feet, I'm estimating. These are mighty and emphatic, assertive, attempting to convey some message, it seems, in the manner of one speaking to a foreigner and trying to drum home sense by extra volume.

I fumble with my digital recorder and so am later able to count an astonishing ninety-five blows in forty-three minutes—sometimes singles, sometimes doubles or triples, never more than four in a string. And whereas two nights ago, there was a certain timidity in the performance, a quality of "off over there," a sense that the communication may not be meant for me, tonight there's no mistaking the urgency and directedness.

I am so stunned that I don't think to knock *back* in any consistent way. Just occasional flailing wallops with my club against the nearest tree. Instead, stupidly, I concentrate on keeping the video camera trained into a lush area to my north, ready to catch dark movement through that wall of full-summer trees.

Eleven days later, I am up to the task of dialogue when, again at dusk, and from the same direction, comes the loud barrage. I answer three with three, and sometimes get three right back at me. But sometimes I mix it up, anteing single knocks after hearing a quadruple, or vice versa. Occasional pauses I let ride, not wishing to seem callow and over-eager, though of course I am.

After fifteen minutes of this sporadic call-and-response, I grab my camera and proceed with it, recording, held down by my belly because, apparently, we're umbilically joined. Still using my club to check in, I also start talking. "I'm not a threat. Ha, that's an understatement! There's absolutely nothing I can do to hurt you. I just want to get closer to you. I'm not a threat . . . *Ha* . . ."

The knocking steadily recedes before me but (importantly) does not quit. "I'm not here to cause a problem for you. Just give me a sign and I won't keep following you. I can't understand what you're trying to tell me. This isn't anywhere near as bad as using my spotlight in the middle of the night, is it? Should I keep walking or go back to my area?"

I go back to my area. I decide to play thirty seconds of the famous "Sierra Sounds" recordings, frightening beast-like snarls, territorial grunts, and other, very different vocals that strike one as a definite form of speech. This audio was obtained by Ron Morehead and Alan Berry in the early 1970s at a remote site in the Sierra Nevada mountains of California, where the team also had brief nighttime sightings and found classic Sasquatch prints.

My conversational partner actually seems to listen, because it's not till I hit STOP (and at that not even *three seconds* after) that a single sharp reply is made, like an acknowledgement. Then nothing.

Until much later, at 10:25, when from an altogether different part of the ravine, behind me to my south and probably a quarter mile away, comes a loud single slam. I gather my wits and produce a single of my own, albeit feeble. To which I get four powerful cracks back, then, again, nothing.

At 1:21 AM, though, while I'm sleeping. the recorder picks up a brisk stick break followed by retreating footsteps, definitely bipedal. They start off the moment the stick cracks, then they cease for eighteen seconds, then they resume, fading out of range.

The Homefront

Since last fall, I've been keeping my eye on a peculiar formation in the woods beside my driveway, just eight feet from where I walk and drive—an obviously constructed grouping of many sticks, all parallel to one another, leaned up at a forty-five-degree angle, from two sides, onto the "main beam" trunk of a tree either fallen or pushed. The whole effect is like a smaller and more primitive version of the T-pee structure in the ravine. Like the proverbial watched pot, though, it doesn't change, season after season.

Add to this, now, a brand new structure, an extremely blatant sapling arch at the base of my driveway, in the scrub across the intersecting dirt road. It appeared sometime in mid-July, shortly after the three gifts. I am sure of this time frame because I've been seeing and regularly dismissing it since then as too good to be true, which demonstrates to me, once again, the stubbornness of resistance, the unbelief resident even in one who's bending his life toward this creature.

When I finally make myself go and examine this thing, on July 28, I find that (Ta-Da!) it is anchored down deliberately with several wooden cross-pieces. The humor suddenly strikes me: if it's a kind of road sign, showing where I live, then what better location than . . . on the side of the road?

The next day, two (human) guests arrive, female friends planning to stay for a week, and write, in the two cabins I have on my property.

Their first afternoon here, as she is walking the path behind my house, the woman from the West Coast is startled by a tree creaking and collapsing nearby. She is able to whirl just in time to see it hit the ground, but nothing of a cause. It broke in two midway up, an old, dead birch, but there was little breeze.

When she reports this mild oddity to us later that day, the other woman, the Vermonter, says, "Oh, that reminds me. I heard a tree fall and crash just a little while ago, too, in the woods that way. But I couldn't see it."

I can't help it. "Hmmm. Well, you know . . ." Neither of them knows of my present obsession. "It's just possible I might know who's to blame . . ."

I wouldn't share so much information if they didn't seem genuinely intrigued and keep asking more questions.

But that night, her first alone in the cabin, the Californian can hardly sleep, hearing things out in the woods. The next night, she sleeps in the house, in a room with an air conditioner blaring white noise, and the day after that, with apologies, she flies back home.

The Vermonter, on the other hand, is into it. Hanging from the basket occupied by a stuffed raccoon, on a tree trunk, she leaves an arrangement of pretty pink flowers. In the morning, not five feet from her cabin door, she finds a little ball of fuzzy yarn, also pink, a close match, sitting in a young pine tree. Of course, she is delighted.

Even stranger is what happens next, and I'm in no position to explain it. I hesitate to include it but I'll do so for what it's worth.

The background: a) This woman has been concerned, for several days, because her period is late; b) She is very fastidious in how she keeps her clothing while on the road, clean clothes always folded, and neither clean nor dirty lying around.

Understandably diffident about sharing this story, over dinner, she feels she must. This afternoon, returning to her (unlocked) cabin after going for a run, she found a pair of her underwear spread on the love seat, crotch exposed, a few brown spots of blood visible. Freaked out, she nevertheless feels she is being told not to worry. And the very next morning, her period arrives.

Late Summer and Fall

I can group these two together because less happens now than back during June and July. Nights in the ravine often come up empty, or give me only ambiguous, far-off thumps and stick breaks, potential footfalls, easy to write off. When I become discouraged, I remind myself of Robert W. Morgan's wise counsel: "It may take months, even years, of repeated visits to your research site in order to produce results." And here, in the infancy of my research process, I've been so lucky, having ample results already.

Yet, really what I want is another visit from Thumper. During that experience, now many weeks past, I can remember assuming that I was, after all, nestled in the midst of a reliable escalation, that interactions would only continue and enhance. But whether it's because of my spotlight escapade, or that the family has relocated, or simply that whoever was smacking the turf like that so frankly and so nearby crossed a line that night and then chose to back off, or was *told* to back off (perhaps he or she was an uppity youth in need of correction), my field work reaps precious few rewards for the remainder of the viable season.

In any case, through August and September, I keep showing up and sleeping in my spot, making my little sounds up there, playing my daughter's hopeful voice. Sometimes my recorder contains, in addition to my snores, possible news of Thumper's re-approach, but again, nothing even 10% as definitive as what I was treated to on June 25.

Leaves fall off, and this makes for lousy conditions, because the sort of deep, inky darkness needed to create safety cannot occur; even beneath cloud cover and a new moon, ambient starlight filters down through the bare branches.

It's not until October 12 that I finally record, once again, a worthy stick break not far from the mic, not far from my inert body, followed by several promising, earthy thumps. I'd likely attribute these sounds to the hooves of a deer if not for the fact that today, on my hike into the ravine, I encounter something truly wonderful—a dramatic tree twist at the top of the logging road entrance, very near where the two young trees were arched across the path fourteen months ago, and exactly where that other tree was brought down over the trail during my first overnight, September 2007.

There is no mistaking it. This tree, eight or ten inches in diameter, is bent at a ninety-degree angle so that it now crosses the entire path, three feet above my head, as I peer up. The twisting action has buckled the structure, popped loose the tree's three constituent, longitudinal segments, so that now it looks like a huge, woody braid. No human hands could have accomplished this feat.

Whatever the inherent meaning of this gesture may be—even if the opposite of my interpretation, even if this is no "gesture" at all but something else—I'll take it as a tremendous pat on the back for me alone, a vote of confidence to tide me over the long winter, dreaming of spring.

As Ammi has told me, "Baby steps all the way, but steps nonetheless." Also, I try to firmly lodge in my mind, again and again, the open-ended fluidity of this whole course of research. In Morgan's words, "It's like a chess game in which nobody wins and no pieces are lost."

And whenever I drive through my little village, past the baseball field where I played Little League a third of a century ago, past the town's general store and post office, its U.S. flag, the feed store, the two dozen houses, past kids playing catch, swinging, shooting baskets, I think, "If they only knew." And then I recognize that even if they knew, most would find a way to ignore it internally; as Texas #2 says, "It goes into the non-thinking part of the brain." Pets and satellite dishes, barbeques and beer, people mowing their lawns or tending their home grounds, resting on front porches, catching up on gossip—and all of this, all of us, not half a mile from a cleft in the earth where another race shares our taproot primate practices, playing and chatting, too, and conducting their business, a whole culture thriving on the graveyard shift.

Newark Daily Advocate

Newark Ohio, Wednesday, August 1, 1863

Man Or Gorilla?
The Extraordinary Character Who Is Scaring Canucks

Ottowa, Ontario, Aug. 1—Pembroke, about one hundred miles north of Ottawa, has a lively sensation in the shape of a wild man eight feet high and covered with hair. His haunts are on Prettis Island, a short distance from the town, and the people are so terrified that no one has dared to venture on the island for several weeks. Two raftsmen named Toughey and Sallman, armed with weapons, plucked up sufficient courage to scour the woods in hope of seeing the monster. About three o'clock in the afternoon their curiosity was rewarded. He emerged from a thicket having in one hand a long stone and in the other a wooden bludgeon. His appearance struck such terror to the hearts of the raftsmen that they made tracks for the boat, which was moored by the beach. The giant followed them, uttering demoniacal yells and gesturing wildly. They had barely time to get into the boat and pull a short distance out into the stream when he hurled the stone after them, striking Toughey in the arm and fracturing it. Sallman fired two shots, but neither took effect, the giant retreating hurriedly at the first sound of firearms. It is more than probable that

the townspeople will arrange an expedition to capture, if possible, what Toughey describes as a man who looks like a gorilla, wandering about in a perfectly nude condition, and, with the exception of the face, completely covered with a thick growth of black hair.

For these newspaper articles, I wish to thank Scott McClean, whose compendium, "Big News Prints," gathers together many hundreds of such historical accounts of human-Sasquatch contacts from the 1700s to the late Twentieth Century. Visit www.McClean.org.

CHAPTER FIVE

Habituation Families One Year Later

In early November 2008, I travel south and spend time on-site with the Texas #1 and Texas #2 families. I also meet up with Ammi from North Carolina, who has been forced to flee her home due to physical abuse by her husband Frank. She was helped to relocate by Texas #2, whom she came to know through the habituators' private Web site.

Prior to this trip, I remained in close touch with all the contributors, and can now pick up their stories where they left off in Chapter Two.

North Carolina

May 14, 2008

We got some bad tornados through this area a few days back—eleven in this part of the state alone that night. We were in bed, and just before a storm hit here, someone outside our window said very loud, "Wah-Coh'-Too!" Just that one word, but it woke up my husband. I hadn't fallen asleep yet, and heard it clearly. It was like it was yelled at the window. It then sounded like someone tried to lift the back window. Frank jumped up, and headed toward the window, and about then, the storm winds hit hard. We both dove into a closet, until it passed. It didn't touch down right here, but just down the road.

It got really scary for a while, but all we lost was some large limbs here.

Frank was getting mad, because I was excited that I finally heard one of their words clearly. I was saying, "Do you think it means, 'Wake up!' or that it is their word for 'tornado'?"

He says, "Who gives a shit, I'm in a friggin' storm, and you are worried about them?"

I said, "Yep, they worried about us, and they are still outside."

That just pissed his puppet more, so I shut up about it.

I still wonder what that word means.

May 18

Had a bit more fun today . . . . a poor soul broke down on the road in front of my house, with a cycle. I have to admit, I took some zoom shots first, from the porch, before I saw how scared he was getting. He actually got off the bike, and tried to hide behind it. Said my dogs were creeping him out, barking at him. He said the barks were coming from the dogs, and echoing behind him . . . poor guy was really scared. Never saw anyone load a bike so quick.

I didn't explain the wooks or the infrasound to him . . . just let him think it was the dogs. Actually, one dog *was* barking, but the rest were just laying there watching him. The one kept barking when I came out. He commented how she seemed to "look right thru him." She was barking at the wooks behind him.

They backed up out of the wheat when I came out to the road, and I fussed at them about scaring him, when he left. I'm sure they got a kick out of it.

I got a "fear wave" of infrasound when I fussed at them, and told them, "Nice try, it don't work on me anymore, I know what it is." It stopped.

It is going to be a strange summer, knowing about them this year.

I searched some online for Indian words. The closest I could find is Apache. "ya-kos'" = "clouds," and "tu'" = "water." Together it would sound very close to what I heard.

[To me:] Best of luck in that ravine tonight, Chris. I still won't camp with them. I'm not that bold here yet at night.

May 22

I found two ripe mulberries on my art easel today . . . . Wow . . . . I went straight to the tree, didn't know they had dropped. The ground was clean under the tree, except for one small pile left there. Guess I'm on rations for those. Can't blame them, I love them too. That explains why they have left my strawberries alone for a few days. Well, at least they are sharing this year. They didn't do anything else at the easel yet . . . . just swiping a bottle of paint every couple days, and leaving me "gifts" . . . . mostly rocks, and a few pinecones.

Even now, I still feel some days that I am no closer to knowing them than when I started. And really I'm not. Nights can still get crazy on a whim, but those have gotten to be more rare than normal here now. There are a few of them who interact, and respond to me here, when they want to, but it is never on demand, anytime I want. They still won't just come out and show themselves, without it being accidental, or as a threat, a brief move. And I still piss them off when I cut the grass, or go into my woods for any reason. They let me know, by disrespecting my house, and hitting on it, climbing up on the roof or porch just to make noise, or irritating the dogs to keep them barking all night.

I can also draw a whizzed-by pinecone, stick, or dirt clod, if I am mowing. So I let hubby mow now. I have pics of him yelling at the woods here, and just staring them down. He has given up on the total unbelief in them existing, and has settled into just not wanting to discuss them at all.

I think I have established that the house is ours. But they still claim the outside, including the outbuildings here. And we have a day/night sharing of the yard. But they want it left alone.

The garden I planted is mine, but they feel free to "trade me" my crops for rocks, sticks, pinecones, feathers, and bits of trash they find, of any kind. Occasionally, I get a dead bird, turtle, frog, or rodent. "Mmmm . . . wookie stew" is all I can say when I find it raided, and those in the place of it. "Wookie stew" is a running joke between me and my hubby, when he asks on the phone from work, "What's for dinner tonight?" Hubby really hates that joke. They still consider all of the old fruit trees and berries here as theirs, and have let me know that.

They do make me smile and laugh at times. Sometimes one will abandon the usual bird and animal calls, for a funky sound, to get my attention. Taking all of my garden tools and stacking them all up together. Lining pinecones in a circle, around my flower beds. I find pinecones in the strangest places here . . . . and that always makes me smile. I guess I like that they care to make me smile and laugh. It makes me feel as if I am gaining some trust with them. Sometimes I wonder if there are just one or two that "get a kick" out of watching the old woman happy, while the rest just still hate me being here at all. One of those things I want to know, but may be better off not knowing.

I have learned to be vocal here with them. I tell them if I like something they have done, and if I don't. It's the mom in me . . . I have to "train everyone" to get along together, like I did when my kids were young. Or like a new relationship and the "control games" to establish the boundaries. Only now, it's me and some people I can't usually see, sometimes can't hear, and whose language I don't know. I think they do understand English. I just don't know the other language they speak. I think it is a very old native language, and like ours, has probably developed its own slang thru time.

Sometimes I enjoy them being here, and other times I wish my life was normal, and I didn't know about them. Normal as in, not discussing or arguing about what the people in our woods are doing, or have done. Normal as in Frank not yelling at me, "STOP INTERACTING WITH THEM!"

There are days when I try to interact with them, and days when I just ignore them. They do the same with me . . . days of interaction, and days you wonder where they are, as they are so quiet. I worry about them in stormy, very cold, and very hot weather.

I wonder how they deal with the bugs, as the biting flies and mosquitoes get bad here. Then I remind myself that they have never done it different, and are probably fine.

The hardest thing is pretending not to hear or notice them, when people who don't know stop in, and they decide to give a mid-day owl/dog bark/chatter and whistle number, from the nearby brush. When it stops your guests in their tracks, and they are staring at the spot it came from, and looking at you, or asking questions you don't want to answer, well it can be funny to me at times. I have a bad habit of laughing at that, and offering no explanation except "Yeah, that was weird." I have lost a few friends over that. But I would rather they think it is this place that is haunted, or strange, than tell them, and them think me crazy. I have done both, so I opt for the "strange place" and just meet them elsewhere, for company.

Finding out that my closest neighbor also knows about them helped a lot.

It was a very awkward conversation start for us both, to learn it. "What do you hear?" "Have you ever heard . . . . seen?" Once we both realized we both knew, it was great. Plus we are both relieved that we have the places on both sides of their sleeping area, so we know they won't be bothered there. We agree that our nights are better spent in our houses. Seems they also have that night ownership in his yard too. It took us five years to have that conversation. Mainly, because he thought my dogs where stealing his chickens. Once I was able to tell him they have been known to "trade" for chickens too, we have gotten along better, though he wasn't aware of the "gift/trade value" of the local pinecones and rocks! He had never actually seen one of my dogs over there, but he had found some strange stuff in his chicken house. It also explained how the "dogs" were getting the coop unlatched. I had lost rabbits from cages here the same way.

May 28

I did my easel thing here. Set it up right at the very edge of the woods, and with my back to them. I painted a pic of one peeking thru leaves, and left it and blank paper out. The first night they swiped (er *traded* for) a bottle of bright metallic blue, for two dead baby snapping turtles, and left paint scattered all over my chair, the ground, and the surrounding leaves.

When I found it the next day, I painted a picture of one of the turtles, and left it there, with some blank paper on the side. That night, they "painted" all over my easel, my pic, the paper next to it, the chair, leaves. Looked like a three-year-old got ahold of it. They gained a bottle of pink, and a sky blue that night. They also scattered my brushes all over the ground, around and under my chair.

The ground around the easel looked like I had eighty people stomping around it for a week, completely mashed and compacted down hard. A new very stomped trail appeared from the woods, leading directly to that spot.

Rain had me bring it up on the porch, where it has been since. I have heard them play around there late at night, and find a bottle or brush moved now and then, but no more painting, nor have they taken anything. They have left a few gifts, like a ripe mulberry on my easel. My kids are grown, so now I have some new "art" to hang on my fridge!

May 31

I went out to water the garden today, and took my camera. I was home alone, and when I got to one side of the house, I decided to "make a rainbow" with the spray, into the sunlight there, partly to amuse myself. So I was spraying away, and snapping pics, and I started thinking about when my granddaughter stayed with me last year, and how she loved to see me make rainbows, and she would play in the spray, running in and out, and singing "Somewhere over the Rainbow" with me. I started singing it out loud, and was doing a silly little dance we used to do together, and didn't notice or hear hubby pull in the driveway from work. He walked up behind me, to see what I was doing, and here I was, just a—spraying the hose in the air, and singing, and dancing away.

He laughed really loud, and it made me jump and stop. He said something about how I had finally just lost it, and started back to get his stuff from the car.

After dinner, I sat down to check the pics. Wow, seems our friends liked the rainbows too!

And the hose. I have woken up many mornings to find the hose on, and have the water bills to prove it. Hubby always blames me for forgetting and leaving it on. Like even when it was spraying all over the porch, towards the front door . . . like I just walked thru the spray, inside, and didn't notice it was on?

The last time, I fussed and yelled at them about not turning it off, and it has been on two times since, but the nozzle has been shut to close the spray.

They also empty my dog tub at night. It holds fifteen gallons, and they will empty it, on hot days, when I have just refreshed it. This winter, I got pics of a five-gallon bucket that went missing here, and showed up in the woods near the pump house, sitting upright, full to the brim with fresh water. I think I busted them trying to carry it off, and they just sat it down and hid. It was on the start of one of the trails that leads to the nest area. I left it there, and it disappeared that night, and the empty bucket showed back up in the pumphouse the next night. So I have set up another tub, near that woods area, and kept it with fresh water. Nope, they never touched it, but they still hit the dog's tub. I figured they trust more what they know I leave for my dogs.

I see them dart across that area a lot, out of the corner of my eye. I think that is what Frank is seeing too. He will turn suddenly, and just stare. That or he got popped with another pine cone, or a stick, for ending the rainbow dance. He usually gets mean when they pop him.

He did go out and take the trimmers to the yard edges, and he really cut up some of the areas they like to hide and watch in. I was kinda mad he did that, but I know they mess with him from those areas, when he mows, so I just figured he is trying to keep them back off him some.

He got ill with me just now when I asked him why such a severe trim job there. Has to be hell at times, being him.

June 5

Hubby came walking with me around, just going on dusk. He was picking with me, because I found a pile of sticks, next to the back of the barn, and I took pics, and took one piece that looked chipped up, and put it in my pocket.

I was pretty surprised he stayed with me, as I went behind the barn looking . . . where he usually won't go. Well it didn't last long, next thing I knew, I heard him make a kind of "hmmmptf" sound, and he turned and was walking fast towards the house.

I had just heard some twigs snapping, and rustling, close, and figured it spooked him. Oh well. I turned back around to look where I had heard it and . . . they just sat there, and let me take two pics. Then I felt strange just taking pics, and I lowered the camera. They turned and left. I stood there a minute, and then went into the house to see if hubby had seen them.

He was sitting on the couch, looking at the TV, and wouldn't even look at me, but he was shook up. I started to ask him if he saw them and he just yelled, "OK . . . I DON'T WANT TO TALK ABOUT IT!! EVER!" And then he went and got in bed. I went in the bedroom, and he shoved my pillows at me, and said, "Just leave me alone tonight, please."

So I am on the couch tonight I guess.

I guess he saw them . . .

They were like in a pile. Little furry ones, kind of stacked in a pyramid shape.

Well, that was a definite first here. I'm still trying to process it, and am wondering why they did that, and if they will do it again. I can't even joke about this one. Not feeling scared, just don't really know what to think about it.

I haven't heard them at all tonight, and the dogs are not up and barking like they usually do. Come to think of it, the dogs usually follow us all over the yard, but today, none of them came behind the barn with us. They followed us up until I found those sticks, and I didn't notice them leave, but none were around when I turned to go back to the house. They were all up on the porch.

Do you think I should leave them a gift there in the morning?

Or should I just ignore it?

I think it did kind of scare me. Not then, but now it kind of does . . . not knowing why they did that.

From a phone conversation with Ammi

—I was talking to [another member of the habituators' forum] and you know what I realized?

—What?

—The other day he had gone and trimmed up those bushes just there, where the little ones showed out yesterday evening, severely clipped those things. I was really torqued with him that he did that. And he got ill with me, talking about, "Oh, you just trying to give them things a place to—" And I said, "Yeah, because that's the way they are, y'know, you gotta give 'em respect." And I said, "That's why you're getting napped with pinecones." And then he says, "I don't believe in that crap!" And I'm like, "Yeah okay, it's the dogs and the deers and the owls throwing that shit . . ."

So when I went outside, I apologized to them. I was talking to the woods, y'know, and I was telling them, "You know I would *never*—'cause I'm the one that usually does the trimming—I would never, ever trim this so severe." And I said, "He's just an idiot. He refuses to believe in you. I know you guys are throwing stuff at him." I said, "I wish you would just step out in front of him."

—Oh, and this was just a few days before the episode?

—Two days before.

—When I was talking to [fellow habituator], I was saying, "Why did they do that?"

And I said, "Well you know he trimmed up that hedge over there, that bush." And as I said that I said, "Oh my God!" Have you ever got that feeling, like when someone hits a certain note in a song, where you get that tingle all the way through your body? Well, that's what it was like, I knew that was it. I got that tingle from my head to my toes.

—So they were in the same place that would've been hidden if he had not trimmed the bushes back in there?

—Yes sir. They sure were. They were crouched down like they were hiding behind the same bush but it wasn't there. And they were in a pyramid, a pile of them.

—What size were they?

—Different. Different sizes.

—Were there two?

—Two? No, there was a pyramid of them. I felt stupid. I still felt stupid this morning, because here all this time I've been feeding these things, trying to interact with them, thinking what you want is an interaction with them, y'know. And what did I do when they finally showed themselves? I took pictures of them like they were some kind of freak show. I didn't say a word to them. I just took two pictures and then stood there realizing how stupid it was, so I just lowered the camera and stood there staring at them. Still didn't know what to say. And when they left, they didn't stand up and turn around or nothing. They just faded backward into shadow, and the foliage. It was pretty deep dusk. You can't make out details in the pictures, just general shapes. It was maybe a minute or less between when I put the camera down and they faded.

—But before that, you got the chance to really feast your eyes on these folks.

—Yeah, and they looked just as confusing as . . . I figured out why so many of the pictures are so confusing. I'm going to be able to pick them out better from now on. They stack up on each other. You got arms and legs and faces all tangled together from a whole bunch of them. That's why it's looking so weird.

—And like the little ones are gripping onto the fur and . . .

—I mean, some of them are laying down, some got their heads stuck in sideways. Some are over from the top. They got their arms over top of their head.

—Ah, just to mess with your eyes?

—Yeah, it's all an eye-screw. And I was like, *that*'s why you can't make out a clear face. One will have his hand cupped over his face, or one of them next to them's got their hand cupped over the other one's face. It's wild, and smart.

June 13

I have always left the blinds open parts of the day, and some of the night, before bed, so they can look in here also. That is probably why if I get up and move around in here for over an hour, and don't raise those blinds, they will start to "tap tap tap" lightly on the windows. They know my reaction is to simply raise the blinds, and go about my business. They know I won't run or jump to try to see them.

I may ease a camera around the corner but my face won't be with it. They are getting me trained pretty well here. They almost have me tamed, habituated.

I also know, if I don't walk around for a few days, and just ignore them, take no pics, etc., they come in closer, and get more vocal, to try to get my attention. They are like children like that. I haven't had them show out again, but I notice in the pics, they are right at the edge.

June 28

Frank was loaded up to take his daughter back to Georgia. She and her boyfriend have been here a week, and besides one tour around our place, have stayed at his mom's house. Frank was in the car, and the boyfriend also . . . . Robyn decided she wanted some apples from my tree, about the same time I came out to say my good-byes.

Boyfriend got out to hug me bye, Robyn was walking back from the apple tree next to the barn, and we all heard a very loud, "Frank!" yelled from behind the barn.

I had to just laugh, it was a very good imitation of my voice.

Robyn *ran* to the car, Frank jumped out glaring at the barn area, Boyfriend and Robyn looked at me confused.

I just kept laughing, and told Frank, "You explain it on the ride." I looked at Robyn and Boyfriend and told them, "They are harmless, and your dad still doesn't believe in them. Ya'll have a nice safe trip home."

And I went inside . . .

* * *

I finally get to spend a couple hours with Ammi in Texas, November 13. After the woman of Texas #2 helped her to relocate, to escape a dire situation back home, and put her up for several weeks (where they compared notes, many a late night, on their habituation projects), Ammi has found a steady living situation just seventeen miles away. Her current place abuts the same type of undeveloped woodland as at Texas #2. And both houses are very near a major river system that boasts a high concentration of sightings.

Not surprisingly, then, this new spot is "active," as well, especially to the trained eye and ear. She points, first, to the roof, where two bricks sit, too far from their source (brick chimneys) to have simply fallen like that.

"The couple that stays here sometimes told me they heard footsteps on the roof, really freaked them out. And they also keep seeing movement in the woods out of the corner of their eyes. They don't stay here much anymore."

On her computer, Ammi shows me an amazing recent picture. She's not had Internet access for a while, so hasn't been able to send me this image. "One evening I heard all this commotion, they were shaking branches in the trees down behind here, and just like back in Carolina, I snapped lots of shots. And when I came in and zoomed it, I said, Oh my Lord. There are some that have less hair on their faces. You can see he's got a wide nose, almost like ape-ish. You can see his mouth real clear. You can even see that line that runs right there from his mouth to his nose. That's the clearest I've got of one . . . his eyes. That has to be the best picture I've took, I think."

Sure enough, I can make out everything she's claiming; it's a distinctive face that seems to hang among leaves.

Next, we go outside and enter the forest just there, fifty yards behind her house. Ammi has been back here only once before, and where she crossed the barbed-wire fence, a tree has now been laid, along with lots of branches, sagging the wires.

"They'll do this. A human comes in their area, they'll do what they can to discourage you from coming back. And make it look as natural as they can. Well, to most people."

Proceeding deeper into the woods, we find many characteristic structures—little sapling bows, larger tree arches, and cases of branches being woven together to form rudimentary enclosures, the ground beneath which often swept smooth, free of brambles and other undergrowth.

I'm impressed by Ammi's capacity to distinguish between natural and constructed formations, agree with her assessments, and generally find her level-headed and just as perceptive and articulate as her many months of emails, and her telephone presence, have conveyed.

Oklahoma

June 28

Last night one of the Locals nearly made an appearance. I was sitting on the swing facing my son-in-law who was dropping off my grandchildren. The yard was set up with the wading pool beside the large pool. The kids had already bailed into the pool. Roger was saying good-bye. Sprout [grandson] was in the wading pool and facing towards the house. It was just before dark. I saw movement at the corner of the house. It was a black shadowy figure that darted around from the back towards the front door. The figure moved further out than necessary before diverting back towards the front door.

This was behind Roger's back. Sprout started pointing and yelling, "Granmaa ook ook!" He took a couple of steps forward watching the area where the shadow had moved through. He got out of the pool, walked a few more steps in that direction, then turned around and came over, crawled up and sat down by me on the swing. Son-in-law was fat dumb and happy and missed the whole thing.

This Local was so dark/black that it even stood out against dusk. I am not sure how to describe it other than it was about the size of a medium-sized adult. Broad shoulders. I am not sure if he was hunched. I got the impression he was ducking as he went under the limb of the pine tree.

Later Dragon Hunter and Fuzzy Wuzzy took some sandwiches out to put in the snack buckets and both were quite unsettled when they came back in. This was about 10:30 pm. Bobbie said that the yard was scary, he saw one of the monkey people out there. He didn't know this one. That for some reason scared him.

Around 11 pm I stepped out for a smoke and on the west side of the house I heard something moving around. It sounded large. It also sounded as if it was picking up a sheet of plywood and dropping it repeatedly. It was not exactly inviting vibes. The geese and dogs were very noticeably quiet.

After the face painting and the hair spray, Dragon Hunter announced that one of the monkey people had been standing off to the north in the bushes watching us. He had seen him. He showed me about where he had been. Same area as last year.

Sprout tried to bolt for the brush a couple of times pointing and saying, "Dah." He never gets loud when he says, "Dah." It's always said quietly. I don't know what's up with that. He can be quite loud when he is playing.

June 29

Today, Sunday, we cooked out. While hubby and my son were talking at the patio Sprout was on the trampoline playing. He stood up, pointed towards the playhouse, and said, "Dah." He looked at me and pointed, saying, "Dah" over and over. He walked to the edge of the trampoline as if to get a better look.

Fuzzy Wuzzy asked him where was Addy [another name he uses] and he pointed towards the north. He looked back towards where he had been looking and seemed disappointed. He went back to playing.

Later, around 7 pm, just before it was time for the kids to leave, Sprout and I walked out into the play yard. I had walked away from him, leaving him sitting on the swing. Fuzzy Wuzzy was talking to me and when I turned around Sprout said, "Ook ook!" pointing towards the gifting stump and habitat pile. I couldn't see anything from where I was standing. I watched his expressions as he looked towards the stump. The best way to describe it is recognition, like when you run into an old friend and you just feel happy to see them. Sprout's eyes light up and he smiles and laughs. When he pointed to Dah earlier and now when sitting on the swing looking towards the stump, he shows happiness from the inside out.

I was going through some of my pictures. Sprout and Prissy Princess was looking at them with me. Both Prissy and Sprout were quick to point out the blobs. Sprout pointed to several, clearly saying Dah. Others he just pointed to and looked at me like who is that.

July 1

While I was on the phone with Furbaby [fellow habituator], Sprout came into the kitchen. Furbaby suggested I ask him if Addy and Dah were outside. He climbed up on his bench and began looking out the window. He stared off towards the habitat pile while looking. That look of recognition in his eyes. I didn't see anything out there. Then it was as if he shut down. He has done this for months. When I begin trying to see what he sees he stops looking.

Today while he was in the play yard Magilla and I asked him, "Where is Addy and Dah?" He stopped playing and looked all around. He walked around the yard looking. He was quiet, then he pointed to the front of the property. He stared hard at the cedars and said, "There." Very plainly said, "There." I had my back to the cedars and was making an effort not to look where he was looking.

I can't prove it but I think a lot of the sapling bows are just children at play. I know what the researchers say. They try to make it so complicated that there are these big mysteries. I think that the little ones play like children do and at times they ride the trees that are bent while they are playing. There are lots of signs that indicate communication, I won't deny that. I just think that little ones use the trees to play in also.

I have several trees that are bowing and up top in adjacent trees the leaves are missing. These are the same type of leaves that Dragon Hunter and Fuzzy Wuzzy have said they see them eating. They hide well in the trees also. They like to watch us from these places. I have also noticed when one tree is bowed to the point I start paying attention to that area then another tree somewhere else starts bowing.

July 3

This is the second time this summer this has happened. Someone short knocks on the front door. Or maybe it is someone taller with a long reach. Twice within the last month someone has knocked on the front door. No one has been there. The first time hubby and Magilla were here and heard it. The first time it happened we thought that Sprout and Fuzzy Wuzzy had arrived. The knock was three then three. Today was a steady knocking on the bottom half of the door. As if someone was trying to get our attention. Magilla and I immediately went out to see what was going on. Nothing absolutely nothing.

August 16

I believe they may have those that are from the wilderness and those that are closer to our back doors. That makes the differences in population, evolution habits, and even the way they look. The ones closer to us are on warp drive adapting to us, continually changing. Those in the wilderness are on a much slower pace.

August 28

This happened Sunday late afternoon/evening. I was out in the yard. I think better when I am outside. Less distractions in the flowerbeds. Anyway I looked up and across the yard a large (probably around 7.5

ft) male was standing just outside of the tree line. He was watching me. It was as if he was waiting to get my attention. When I noticed him I watched for a couple of seconds and could even see the white teeth. It was not actually a shock, more of a surprise. I blinked and was wiping sweat from my face. He disappeared. It only lasted a few seconds.

My impression was he wanted me to know that he had come for his family. They have now gone. I don't know where. I feel better knowing that the female didn't leave alone with little ones. Nothing I can put my finger on. It was as if he was letting me know he had come for his female. He was very black in color. His hair around his head was shoulder length. The body was covered in black hair. His hair appeared short and uniform. Well groomed. His head was round not like a gorilla. Not a huge bulky male more tall and slender. I checked the area where he had been. He had actually stepped from the woods into the yard, there were impressions in the grass.

About three weeks ago, I sent Fuzzy Wuzzy out to turn on the lights around the patio. He came back in and indicated that it was very creepy out. I finally pinned him down. He said he saw a new guy in the yard. Around 9 ft tall. He looked funny because his head was a different color than his body. Fuzzy used the trees this one was near to measure the height. He says it's too weird to explain, it was like meeting a stranger in the yard. He said when it's the ones that stay around more it's just like "Oh hi."

September 21

I have some odd things and (as hubby says) "new hillbillies" around. They are just different. They are more tense when they come through here. I believe that Addy left with that male. She may be back in the area but not like she was . . . . I can't put my finger on things to say for certain, but there is some kind of change or disturbance in patterns. My locals that were comfortable around here seem to be less noticeable. It could be just me also. Addy, Dah, and the other little guys were around a lot. I would see flashes. Sprout was quick to point them out. Now he isn't pointing them out. He looks for them. He has pointed to a couple of shadows. When asked about them, he don't seem to know them. Not like Addy and Dah. Kinzi says she has seen several that look more monkey than monkey kid. She don't seem to know them either.

On September 24, while crossing the street, Sprout and his mother were struck by a car, and the boy suffered several injuries, including a skull fracture.

October 28

Since Sprout's accident he has not rested very well. He often dreams of the accident. Or I believe he is dreaming about it. He screams in his sleep Help me Help me. Stop it Stop it. During these dreams he is restless. I have to physically restrain him at times. Often it is hard to wake him up. Once he is awake he don't want to sleep again.

On Thursday Oct. 23, he had finally gone to sleep around 11 pm. Around midnight the dreams began. Not able to settle him down I woke him up. After a few minutes I changed the tv to a public tv children's channel and he half watched it, while complaining about his tummy. Then he became excited looking out the front window into the dark (not unusual). He started saying Henry. Henry's here. He pointed out the window and jabbered to the dark and Henry. After doing this. Sprout hopped down and went to the door. He tried to open the door. During this time he was saying, Henry come in the house. Come on in, Henry. He invited Henry to come in several times. I sat watching this, thinking I hope Henry isn't heavy and breaks the porch. A minute or two later Sprout climbed back up to the window looking out. He didn't mention Henry again.

The next night Sprout was again restless and having bad dreams. Around 11 pm I began to hear a humming/singing outside. Really.

It was a low sound. As odd as this sounds Sprout settled down and seemed to sleep peacefully. I don't know how long the humming went on. In the wee hours of the morning Sprout became restless again. I barely recall holding him and the humming beginning again. This was the third time I have heard the humming now. There has not been any wind when I hear the humming. Or that is to say I don't observe the trees swaying in the wind. I have investigated around the house looking for some sort of device that the wind could blow through to make the sound, should somehow I have missed noticing the wind. I have not found anything. I can say if it's a hairless person humming they are either bundled up or freezing their butts off. I have not attempted to record anything in a while. I have debated each time I hear the humming on trying to sneak the recorder out. I have decided against this each time. I somehow suspect the humming will come to an end.

October 30

This evening Sprout came out on the porch. He flipped on the porch light and called to Dah. He called several times softly. Then he yelled it very loudly a half a dozen more times or so. He acted very excited while calling for him. Then he went back into the house. A couple of minutes later he came busting out the door a second time and said, KeeKee! KeeKee's here! He looked around as if expecting her to step up to the porch. He went and got a sack of cat food and proceeded to pour a pile on the porch. After making a mess of the cat food he picked it up and put it into a bowl and proceeded to tell the cats they couldn't have it.

New York State

After my last visit to Jerry's encampment, in December 2007, when we heard many wood knocks, both near and distant, he has experienced a long lull. "It's been a year since our last trip together," he says, "and just about as long since I've had good contact with them. They need their comfort zone from us, you know. They must have found a good spot out there. I would too. Deeper in the woods means more comfort for them. They probably have very good reasons why they do what they do."

At first, he suspected this was caused by my visit itself, that bringing a stranger in, altering the whole dynamic from a single-person face-off to a virtual *group* situation, may have broken the tacit contract.

On the other hand, he also takes a broader view: "It seems they move around through the year. If they find a nice new coy spot out there well they'll stay for a spell. Remember you're just right yonder over the next hill. Ya just gotta go out there to have a good time in the woods, and what ever else happens is a bonus. They are a wiley breed to begin with. They just might be honing their skills with us or something, like a game.

"To tell you the truth, the more I think about it I don't think they're disappearing on purpose. They go on hunting trips and find other nice spots in the woods to settle down for a few seasons. Plus my guys have the little one to think of, and his safety is #1. The

Adirondack mountains aren't far from here and that's big woods up there."

His words help me to feel less disappointed by the thinning of activity near me, from mid-summer on. This is not, after all, like studying life within a pond; even in a ravine, there are no such banks.

Texas #1

November 10 and 11, in exchange for one log-cabin-shaped tin of genuine Vermont maple syrup, I am taken in warmly by this family, the parents and their two children, fifteen-year-old Rachel and her thirteen-year-old brother, whom we'll call Mobley, because he tends not to wear shoes outside, running through pastures and woods. These are bright, handsome teenagers, only too eager to share their experiences with me.

In fact, both kids go barefoot this afternoon, and so do I, as we cross a vast cornfield. The stalks have been cut and ploughed under, so what's left is a wet mudflat probably two-thirds of a mile from tree line to tree line. Shoes or boots would be sucked off immediately; our feet sink six inches each slimy step.

Rachel's leading us to "the village," a site famous within the family, a sylvan spot where "the Foots" have sculpted huts out of bushes and low trees, bending and weaving them together into a fine mesh, half a dozen green domes with small entrances down by the ground. I've seen pictures.

"I used to come here in the summer, sit in the huts with my friends. It's so peaceful inside there. I went maybe twenty times."

It takes her a while to pinpoint the location, though, because things here have been changed around dramatically; the village is largely destroyed, leaving just a few tattered remnants of the former structures. Rachel is saddened, thinks maybe she visited too much, making the outpost no longer so desirable.

In case they've rebuilt somewhere nearby, we poke around. And sure enough, after fifteen minutes, Mobley finds a large structure, a sort of "blind" that one can walk into, sit inside on a matted floor that he pronounces "really comfy." This place is formed, up top, by a thick tree that's been obviously broken and bent back downward at a forty-five-degree angle to the earth. The resulting screen of leaves and branches has been augmented by many vines and further branches, stuck and twisted in.

And it makes for a peculiar phenomenon, just like the "two-way mirror" effect Ammi noted in North Carolina, where she could see nothing past the wall of foliage at the edge of her backyard, whereas when she'd enter the brush, her own home and grounds are very plainly visible.

Mobley and I stand inside and watch Rachel's red shirt, moving. "Can you see us?"

"Not at all!"

We exit, she enters, and promptly vanishes, confirming, "I can see you *both*!"

Back at the house, thoroughly mud-covered, we survey the two dozen cages, stacked in rows, now standing empty, many of them badly busted.

"They took some rabbits," Rachel tells me, "and they broke some rabbits' backs."

"Didn't that piss you off!?"

"Oh, I knew they didn't mean anything *by* it. Sometimes, they'd just move them around. And the rabbits are really sensitive. Like even if I were to hold them and make a loud noise, they'd have a heart attack and die. So something that big coming through and opening the cages and moving them around . . . they'd either have a heart attack and die or they'd be put in too roughly and get broken backs and die."

"Why'd they do it?"

"They did it because they watched *me* moving the rabbits around. I knew they were watching me. I could see them sometimes watching me. I'd go out and take care of the rabbits every day. And I'd move them around or whatnot, make sure everybody was where they were supposed to be. And then eventually *they* started moving them around, or trying to. The cages would be all smashed up."

That night, over coffee in the kitchen, she fills me in on some background I haven't heard.

"I was out taking care of my rabbits, one evening, and I glanced back at the wood pile. There was one sitting with his back up against the wood pile, and he had his leg crossed over. I watched him and he watched me. I just stared at him for a long time. He just sat there. He had a gorilla-looking face, definitely. I ran in to go get a flashlight and by the time I got back he was gone."

Her mother joins us. "Remember," she asks her daughter, "that one night we were sitting out in the yard, and having a cookout, and they were hollering and whooping, from the north of us, the south, east and west?"

"Mmmm hmmmm."

"That was early on in us discovering things about them. We were having a lot of cookouts, and they'd be talking to us. We'd whistle and they'd talk back, whistling and whooping."

"We were sitting there on the yard that night and we could see their eyes shining in the trees."

"Green and yellow eyes . . ."

"And blue . . ."

"And one night," her mother recalls, "we were sitting watching TV, and we heard somebody running across the *roof*, Boom Boom Boom, and then it sounded like they fell—CRASH! And then you could hear them roll down the roof and hit the ground. But by the time we got outside, they were gone. They ran around up there more than once, but we've only heard them fall that once."

"Was it funny?" I ask.

"Yeah, we were laughing! Somebody fell!"

She goes on to tell me about a spy-structure that appeared, over a few weeks, at the back of their property. "They watch the kids on the trampoline, they sit back there." The next morning, I see what she's talking about, and film it. "They made a bench, where there's wood in front so they can see through it, and it's tall, and they sit on it, and they can watch the kids at the house. They made an actual *bench*. They moved the wood around, piled it up. An observation post!"

Also, the next morning, before I have to leave, to visit Texas #2, Rachel shows me where, in her room, they used to come and pop the plastic cover off the doggie door, and on her window, in the lower right-hand corner. She imitates with her fingernails . . . tap tap tap tap.

"Trying to wake me up, trying to get some reaction out of me. And they'd smack the wall, too, right here. Late at night."

"Would it make you feel good, like, Oh, my friends are back?"

"No! I'm like, I'm *sleeping*, leave me alone!"

Texas #2

June 17

I came inside from mowing, cooled down, took a bath, and relaxed a bit. Just as the sun was beginning to set, I was talking to [a friend] on the phone and looked back by the goat house—OMG! There was a young BF! It was about 5-5 ½ feet tall, medium build, and even had a *neck*! I don't know if it realized I spotted it, or if it was in a hurry. It made the distance from the goat house to the road in about two seconds. I was *stunned* at how fast this creature ran! Now I know how my husband could have mistaken a young BF for Allison during that rock fight. With the exception of how fast and agile the creature was, it was remarkably like Allison in proportion.

June 19

This evening, about 6:30, I let the dogs out to run a bit and go to the bathroom. I'd been outside about ten minutes when I heard a squalling baby animal coming from near the back fence. The shortest and easiest route to where the racket was coming from was to head west on [road name]. Initially, I thought a coyote had gotten one of my cats. I ran down the asphalt and got about thirty feet from the back fence when a doe busted out of the woods from behind my property, whirled around to her right and stood staring at the woods where she'd just left. I could still hear the animal squalling in panic/pain, and got within about fifteen feet of the doe before she realized I was there. That *should* have been a warning to me that it wasn't a coyote, but I was more focused on the cry from the woods.

In hindsight, it sounded much like one of the baby goats we used to have, that's probably why I was intent on rescuing the baby from the jaws of the coyote. The doe spotted me, then spun and headed away from the squalling young one. I guess the predator and then the addition of a human being was too much for her to deal with. One of the chihuahuas ran ahead of me into the woods, then came running back and stood on the road.

I ran into the woods, across the rocky ground covered by dry leaves, making quite a racket as I went. As I got within about twenty feet (still at a dead run) I hollered out as loudly as I could, "HEY!" Well, instead of a coyote scurrying off into the woods, I was met with a very loud, very intimidating AAAAAAAAUUUUUUUUUGGGG GGGG GGHHHHHH!" I froze. I was too petrified to move. I was located about three or four feet above the creature in the woods. Even at that elevation, the hairy creature was nearly at eye level with me. I could see the top of the head, the left shoulder and a very large muscular left arm. The rest was behind brush. Thankfully, the being in the woods took the still squalling young deer and headed into the ravine. About twenty feet from where the confrontation occurred, the fawn quit making any noise.

By this time, I realized I had been within just a very few feet of one of the Ancient Ones, and a very large, mature one at that. I was shaking all over, scared half to death, my heart was racing and my most urgent thought was to get my stupid butt away from there before the deer killer changed his mind and came back to thump me.

I got back to the house and had a difficult time climbing the four steps to the front porch I was shaking so badly. I immediately went to the bathroom and threw up, then sat on the porcelain throne and found some relief. It was nearly an hour and a half before I was able to compose myself enough to call [friends on the phone]. The

Ancient One in the woods beat a hasty retreat when he/she realized I was *not* doe coming back for her fawn, most likely a bit embarrassed that a human had gotten so close. Had it chosen to attack, I would not have been able to even move. [My friend] said that the poor, scared critter was probably still sitting off in the woods shaking and trying to regain composure from such a scary incident. I should be ashamed of myself! [He] didn't seem to think I should worry much about any retaliation. He said just behave as normally as you have in the past, and don't let the big critter get the idea it had scared me. It's possible that if it thought I had gotten intimidated, then it could try it again on me, or even someone else. If I go through the motions of things being usual, then the critter will just be more careful next time it goes hunting and the prey makes a fuss. And, since I didn't keep him from catching and keeping his meal, there was no harm/no foul (so to speak).

I had suspected the ravine area was used for catching prey, but never realized it was used during daylight hours, too. It *was* getting toward evening, but the sun was still definitely up in the sky, it was basically broad daylight even in the woods. Several weeks ago I recorded the "Whistler" back in the ravine well after dark. The Whistler came steadily up the ravine, accompanied by random wood knocks.

I estimate tonight's hunter to be about eight foot tall. He was dark, kind of a chocolate color, with a semi-peaked head. Not a huge crown like a large gorilla has, about half as large. The hair was between three and four inches long, but longer on the forearm, probably nearly six inches. I couldn't tell due to the amount of brush between us (thank GOD) whether it was male or female. To be honest, it didn't matter! I was just very thankful to still be alive. I didn't notice any odor, either.

Now it's almost 11:30 pm, nearly five hours since the encounter. I'm still a bit shaken, but thanks to the calm conversation with friends, I'm doing much better. I have learned another valuable lesson, and survived to tell others. Or, I dodged another bullet, depending on how you look at it.

June 29

The only "retaliation" I have received was finding all my lawn chairs in the far backyard. I laughed, brought the five chairs back to the front. The next night, the chairs in addition to some of my flower pots and all four of my antique milk cans were moved to the far backyard. Again I laughed, but said, "Okay, that should make us even now because it is harder for me to move all this stuff back to where it belongs than it was for you to bring it here." I haven't had anything else relocated.

I have to admit, this new reality I'm living is an entirely new world, much like Alice in Wonderland where nothing is what it seems. Could the Ancient One world be similar to our world where there are "civilized" AOs and some AOs a bit wild and opportunistic? My thinking is: There are many more AOs than ever estimated by humans. These beings don't exactly put out mailboxes or construct houses. *If* women were being attacked by these beings, I truly believe it would be on the Internet. I believe in my heart these are a society and culture that is very well disciplined. It would not surprise me at all to find out these beings are much more civilized than we hairless human beings are. I have no fear at all of being raped or pillaged when I'm out alone. Even when I first "discovered" them, my fear was more of being the main course of their dinner . . . definitely not sexual entertainment. For some reason, it is the *men* who seem fixated on this aspect of the AO interaction. I think these men are missing the whole point. I believe with all my heart that the interaction has much deeper spiritual implications.

The government does not want to admit these creatures/beings exist. They'd have to admit they have absolutely no control over their behavior. If forced to "protect" the Bigfoot, then the government would have all sorts of lawsuits requesting restitution for lost livestock, property damage, etc. If it is found this creature is a human-type being, then that poses a whole different hornet's nest. Then they've got this radical society to deal with. These creatures don't wear clothing, don't pay taxes, don't get their children vaccinated, or put them in school. They have not bowed down to the authority of the U.S. government, and I doubt they ever will or could be forced to do so. You know what the government and the settlers did to the Native American Indian. God knows, it was a great effort to wipe them out. And it nearly succeeded. That is what could happen to these beings. By "outing" them [to research organizations that can't be trusted to keep one's location secret], you could be forcing the government to deal with a very delicate problem.

I almost feel I am a "Nazi sympathizer" turning in information on the location and behavior of the Jews. That is how seriously I feel about the information I pass on. The more familiar I become with the Ancient Ones' behavior, vocalizations, social structure and annual "celebrations," the more I feel I am being entrusted with what should be *private knowledge.* It may be that *everyone* is not supposed to know they exist. I think those who are drawn to investigate these beings should ask themselves *why.* Some folks always like a good mystery. Some folks have had a sighting and just want to get all the information possible. And some folks like me are *thrust* into the center of this whole thing. I don't believe it was an accident.

July 6

The Ancient Ones have helped me discover a part of me that was missing. This new reality is quite a bit like going back to childhood and discovering the wonders of the world again. I get to re-look at everything with different eyes. If it weren't for the ladies on the forum, I would never be as far along as I am now, and I am eternally thankful. They helped me to see that the great surge of activity I was experiencing here last spring was not a siege but a celebration, connected in some way with the ancient spring ritual called Beltane, Spring Equinox.

What fun it is to go outside and find yard ornaments rearranged during the night! When I misplace some tool, I can now blame the hairy neighbors rather than my own mind! I am finally discovering joy in my life again, and it's been gone a long, long time. It's satisfying to know about the existence of these beings—and no one else in the area has a clue. It's almost like having a secret life: on the surface I look like your typical old woman, living a meager life, eking out an existence in the middle of nowhere, but if the truth be known, I have a very spiritually fulfilling, adventure-strewn, busy, by no means boring life. Pretty cool!

I don't have a clue why these Ancient Ones have revealed themselves to me, but life would truly be lonesome here otherwise. Yes, I have the four dogs and the three cats to keep me company, but nothing compares to a good mystery needing solved. This alter-reality keeps my mind working overtime. I find myself more alert while driving. I tackle plumbing problems myself rather than invite someone over who also might discover my forest secret. And even if they didn't, I don't want the interruption! At night while I'm at the computer, I hear bumping on the rear wall of the house. It's rather comforting to know they are around.

And that is another change I have discovered in myself—I am more observant of the activities and sounds around me. Instead of slogging through my daily projects, when I rest, I look and listen. I watch what the dogs are doing, where they are looking. I look for the cats to see if they are dozing, or if they are on guard and alert, or if they are chasing each other and energetic. I listen to the sounds of the night. I used to be so closed up to these simple pleasures, now they are mine in abundance. Ancient Ones are amazing mimics. I've heard a brook in my back window—a brook!! I could just imagine a bubbling brook feet outside my kitchen window . . . what a pleasant thought! Not long ago I heard a cow in my backyard about midnight.

There are cows in the area, but close examination the next morning revealed no hoofed creatures back there.

July 19

I ran into one of the lady neighbors as I was walking up the road. She asked what I was doing out in the heat, then said, "OHHH! I know—you were looking for Bigfoot signs, weren't you?" I confessed, then she asked where it was I had my initial sighting, and I told her. Okay, she lives about a half mile northeast of me. I asked her if she had looked around her pond and she said, "Oh, we don't have those things around our place. The dogs keep them run off." I just smiled, said I sure hoped so, then went on walking to the house. Folks certainly have funny ideas about where these hairy woods folks stay and where they go.

July 21

Well, I just got zapped with infrasound *again.* I was trying to put a recorder outside the house, on the north steps. Thinking I'd be kind of slick, I acted like I was moving the bowl of cat food away from ants and relocating a window screen (I'm scraping paint off windows). I moved the bowls of cat food successfully, but when I went back to move the window screen it had gotten darker. I turned on the flashlight and a glint of light caught my eye. I was on the north side of the house, facing the west. The glint was to the northwest about forty feet. I used the flashlight and walked toward the light and realized it was a pane of broken glass leaning against the base of a big tree. The pane of glass was about an inch thick, and I was seeing the reflection on the edge of the glass. At this point, I heard the "white-tailed deer bark" about twelve to fifteen feet *up the tree* I had just illuminated with the flashlight. Knowing this was no deer, but also knowing the AOs rarely attack, I backtracked and picked up the window screen, but for some reason I decided I wanted to peek around to the back side of the house. While holding the window screen I used the flashlight to light up a portion of the backyard which is enclosed with fencing. When I did that, I immediately got hit with infrasound.

As I've said, I was familiar with the heart pounding, feeling of fear, heightened anxiety. I heard no accompanying growl or bark, I just felt the pulse of the infrasound. This "pulse" was also picked up on the window screen; I was able to feel it with my fingertips. I decided I had pushed my luck as far as I dared, placed the screen on the steps, and left the area, going back inside the house.

The effects of this session of infrasound so far have lasted about forty-five minutes. I am experiencing nausea, rapid heartbeat, anxiety, fear, tremors and a dull headache. I am still having difficulty typing and collecting my thoughts in a rational, coherent order. The side effects do appear to be subsiding.

August 1

I went into the backyard to cut down some fencing. The atmosphere was a bit anxious. I started cutting the wires holding up the gate and got a rock thrown near me. I had come prepared. I said in the most authoritative voice I could muster, "Okay! I just want you to be aware that I have a camera *and* a digital recorder with me, and I know how to use them!"

I hear this very hesitant voice from across the ravine. "Ummm, lady? Are you okay over there? Do you need some help?" Oh man, I felt like a dingbat . . . Feeling rather sheepish, I hollered back, "No, I think I've got things under control now, but thanks for the concern!" If the folks in white coats ever question him, I'm sure he'll testify against me.

While I was out there later putting out food for them, [a seventeen-year-old neighbor boy] came by on the road and began with small talk. Then he got to the point of asking what I was putting out the fruit for. I told them, "Well, just about whatever *wants* it,

I guess." I could tell this boy was a bit uncomfortable, so I said, "Young'un, what's eating you?" He said that he'd *heard* I had Bigfoot on my property, to which I responded, "Kid, they're all over [this part of] Texas. I don't have a corner on the market here." He asked me what he needed to look for to see if they were near his house, so I told him about the tree bends, the X's and the wood knocks. I told him they are not aggressive, but that if you happen to get close to one of their little ones, they use something like a real low growl that we can't hear to make us feel spooked. His eyes lit up and he said, "Oh man! That's what happened that night." And he went on to tell about getting scared when walking home. I told him never to shoot one unless his life absolutely depended on it, because they never travel alone, and getting shot they take *real* personal.

August 12

While I was scraping the overspray off the outside windows, I was able to catch one of the AOs observing me from the brushy area between the house and the county road. I was able to view the being without spooking it off. The reflection in the window wasn't all that good, though, because I had to keep working and vibrating the glass, otherwise I was concerned the AO would catch on to what I was doing. From what I could observe, it was crouched down in a squatting position, the head almost even with the back of the lawn furniture. Even the face had short hair on it. How they can sit there so very still for such a long time is remarkable. If I squatted like that and then had to move quickly, I couldn't do it.

* * *

I arrive on a mild, sunny late afternoon, Texas autumn, and, accepting my crock of, yes, genuine Vermont maple syrup, she introduces me around to her two adult cats, two kittens, and five dogs, Chihuahuas, a sheltie/rat terrier mix, and of course LuLu, the coonhound whom I'm especially glad to wrestle on the grass—famous, to me, from her role in this woman's arresting accounts.

I'll call her No-Bite, the name her Ancient Ones gave her last spring, after she assured me, "I'm not going to bite you."

We sit out on her lawn, chatting, getting acquainted in person after so much long-distance communication. She's a sweet, kind-faced soul who has been cooking a beef brisket for me all day. As evening descends, one of her neighbors can be heard above us, taking his paraglider up for a spin at several hundred feet.

"Oh, there he goes again. I don't even know who it is, but he's been flying the last couple weeks. Always at dusk."

As the single-person craft approaches, we can hear, from down in the ravine behind her house, five distinct wood knocks, separated by intervals of distance and time. No-Bite's ravine is a lot shallower than mine but still, at seventy-five feet deep and a quarter mile long, and filled with brambles, downed trees, bowed trees, it proves formidable to traverse. Also, the whole affair proceeds slightly uphill, and it's in this direction that the coyote and Sasquatch drive their game—hogs and deer—through. At its outlet, by the road, is where the hunt was interrupted several months ago.

"Hear those knocks?" she says. "They're warning each other to hide, to hunker down. The first time that man flew over, they made a terrible commotion, now it's gotten more subtle." I'm delighted.

After dinner, we sit out in the chill, wrapped in blankets. Indeed, I'm able to spend a total of five nights and days here, and not, for the most part, seeking to gather evidence (like the avid researchers who have visited before, alienating No-Bite with their domineering ways) but rather simply getting a feel for the place, communing with it, exploring the forest and sitting attentive in the yard. Occasionally, at night, we make wood-knock overtures and, I'd say a quarter to a third of the time, either right away or minutes later, receive a crisp response knock from the ravine or from the thick pine forest opposite.

With her permission I do put out my audio recorder, overnights. Once, hung from the goat house roof, it picks up some light slaps against the walls. The next night, though, set beside a bowl of plums left for them, the mic captures nothing so moderate but, at 1:15, an ear-splitting, high-pitched SMACK!, as of one barkless branch or limb connecting with another, in no uncertain terms.

On one of our treks together, No-Bite leads me to a pond within the pine forest that has often seemed the source of middle-of-the-night vocals. The best thing we find here hits me as a work of art: four separate tree arches all in the same spot, curving elegantly and pinned, two of them pinned such that the trees can *keep* curving past semi-circle and into loop. Really just breathtaking. And unlike my own far-northern arches, these cannot be explained away as a by-product of snow and ice build-up.

Another time, she shows me where two long, straight trees have been transported from where they grew, de-limbed, de-barked, and leaned at a steep angle to cross one another twenty feet up in the crook of an ordinary tree. Reminiscent of telephone poles, they're much *more* reminiscent of the elaborate structure we found on the BFRO expedition in Ohio, 2006, the only difference being that, there, *seven* trees were thus stripped, raised, and top-crossed.

And then there is the night of the bonfire in the backyard, which I have mentioned before. No-Bite's twenty-year-old daughter and the daughter's best friend are here with us, and we're roasting hot dogs and marshmallows, having a good time, the two young women acting like kids, giggling and roughhousing.

I leave the firelight and scan the woods with the thermal imager, looking for an upright figure hiding behind a tree, spying on the revelers, its body heat standing out brightly amid the surrounding (cooler) grays. No such luck. But what's that horizontal bar of light, down near the ground, sticking out from behind the debris pile, a perfect vantage point on the bonfire? Should I approach and investigate, shooting thermal footage? Nah, it's probably nothing . . .

* * *

Written by No-Bite to a member of a regional Sasquatch research group

As I said before, I don't have anything to hide. Right now the trail camera you installed is just sitting out there, I don't know if it has photos or not. Bob, what bothers me about your organization is the mindset that you feel compelled to collect "evidence." If the Ancient Ones want to *offer* evidence, well that's a whole different ballgame. Right now, I am quite content just to be along for the show.

I've already given you the resources to take this same journey. I've told you how, I've explained what to expect, and that there are no guarantees. I have absolutely *no idea* why some are accepted, some seem to be chosen, and some folks just sit there and get a numb rear end. It's a strange world out there! And the funny part about it? I laugh now! I haven't laughed like this, this deep belly laughing, in years. I had forgotten what it is like to just cut loose, turn my cares over to a Higher Power and relax. How can you forget how to laugh, a basic human response?

Bob, you know what these people in the woods require of us? *Nothing.* Not a single thing, they just are reaching out because they desire friendship. The reason I was in awe of [two recent human guests] was they were unafraid of these woolyboogers in the dark. I now know this is because these are people, not woolyboogers. I'm cautious, because I'm still new and also I'm alone here, but I know there is no need to be afraid of the hairy folks here. I just feel so very blessed to have had my eyes opened so I could feel how *full* my life really is.

I don't care whether science is ever able to prove these people exist. At this point, I have even given up trying to prove to you that I have them here. I don't need proof anymore. Whether or not they are here is so very insignificant when the big picture is considered. I so wanted you to put down your damned game cams, digital cameras, digital recorders, infrared cameras, motion detecting equipment and all the God-blessed *stuff* you constantly tinker with so that *you* could experience a whole new dimension in "research." What do I know—I'm just someone who happened to be blessed with seeing these reclusive beings. These aren't "chance sightings," they are deliberate. These are an indigenous people. They have their own laws to be obeyed, their own culture, their own beliefs, and social order. Finding these people in my own backyard is akin to finding gold! I don't want them dissected, photographed and ID'd, fingerprinted, probed and grilled. I feel they are now my extended family, and when folks mess with family, I get angry and protective.

How many others are experiencing the same things? I'm not certain. Maybe (and this is a guess) several thousand around the world. What's happening here is also happening elsewhere. If I was the only one experiencing these situations and seeing these occurrences, I truly would have to wonder about my mental well-being.

There are those on the fringe of this knowledge that so want to be included, but for whatever reason they can't or don't see. Many individuals who are still intent on collecting evidence and documenting sightings quickly back off when the deeper, more meaningful experiences begin to happen. That's why many of the folks who report sightings soon drop off the radar.

The research approach presently used is demeaning and insulting, and that approach has been abandoned for a more person-to-person approach. Just like if you were trying to get to know your next door neighbor. Once these reclusive people realize you *know* the bluff charges and vocalizations, etc., are basically a hoax, you end up on a fast track to a whole "other world," is the best words I can come up with. The individual has both feet on the ground, but his/her mind becomes open to something that just defies scientific explanation. If you would like to speak with those taking me on this new course, you'll have to get in line behind me—because so far there has been no spoken word. I do know I have to constantly fight back the fear of the unknown and persevere if I want to go further.

You are wanting blood and guts, rock hard evidence of these people. If I *did* have it, I wouldn't provide it to researchers who want to tag and label. I would absolutely love to think you would drop any preconceived notions and just come and sit and visit, but for whatever reason, God has created you with blinders. Maybe you are not supposed to see what is happening.

Bob, my entire perception of "my world" has changed, and I'm just beginning to understand some of the things going on. The only thing I can compare it to is being mentally reborn. Nearly everything I *thought* I knew has been given a different slant. Once I dropped what I *expected* of these people, an entire culture was opened up to me. Because I was so fearful, these beings spoonfed me a little at a time. What I am able to tell you so far is this: These are a people who do *not* want to be "found." A few open-minded people around the world have been blessed with the opportunity to step through the established wall these people have built. These people fit in with humans like "gears." What humans lack, these people possess. What these people lack, humans possess. Why is it that more and more individuals are having sightings? Because I believe the walls are coming down between the two cultures, but not completely. At this point I feel it is my purpose to plant the seed within you.